Olfa Kanoun (Ed.)
Impedance Spectroscopy

Also of Interest

Progress Reports on Impedance Spectroscopy
O. Kanoun (Ed.), 2016
ISBN 978-3-11-044756-9, e-ISBN 978-3-11-044982-2,
e-ISBN (EPUB) 978-3-11-044767-5

Energy Harvesting for Wireless Sensor Networks
O. Kanoun (Ed.), 2018
ISBN 978-3-11-044368-4, e-ISBN 978-3-11-044505-3,
e-ISBN (EPUB) 978-3-11-043611-2

Series: Advances in Signals, Systems and Devices

Systems, Automation, and Control
F. Derbel, N. Derbel, O. Kanoun (Eds.), 2017
ISBN 978-3-11-046821-2, e-ISBN 978-3-11-047046-8,
e-ISBN (EPUB) 978-3-11-046850-2, Set-ISBN 978-3-11-047047-5

Sensors, Circuits and Instrumentation Systems
O. Kanoun, F. Derbel, N. Derbel, (Eds.), 2017
ISBN 978-3-11-044619-7, e-ISBN 978-3-11-044837-5,
e-ISBN (EPUB) 978-3-11-044624-1, Set-ISBN 978-3-11-044838-2

Power Systems & Smart Energies
F. Derbel, N. Derbel, O. Kanoun (Eds.), 2017
ISBN 978-3-11-046820-5, e-ISBN 978-3-11-047052-9,
e-ISBN (EPUB) 978-3-11-044628-9, Set-ISBN 978-3-11-047053-6

Communication, Signal Processing & Information Technology
F. Derbel, O. Kanoun, N. Derbel (Eds.), 2017
ISBN 978-3-11-046822-9, e-ISBN 978-3-11-047038-3,
e-ISBN (EPUB) 978-3-11-046841-0, Set-ISBN 978-3-11-047039-0

Impedance Spectroscopy

Advanced Applications: Battery Research, Bioimpedance, System Design

Edited by
Olfa Kanoun

DE GRUYTER

Editor
Prof. Olfa Kanoun
Chemnitz University of Technology
Faculty of Electrical Engineering and Information
Technology
Reichenhainer Str. 70
09126 Chemnitz, Germany
Olfa.Kanoun@etit.tu-chemnitz.de

ISBN 978-3-11-055712-1
e-ISBN (PDF) 978-3-11-055892-0
e-ISBN (EPUB) 978-3-11-055716-9

Library of Congress Cataloging-in-Publication Data
Names: Kanoun, Olfa, editor.
Title: Impedance spectroscopy advanced applications : battery research, bioimpedance, system design / edited by/Herausgegeben von Olfa Kanoun.
Description: Berlin ; Boston : Walter de Gruyter, [2018] | Includes bibliographical references and index.
Identifiers: LCCN 2018030273 (print) | LCCN 2018036520 (ebook) | ISBN 9783110558920 (electronic Portable Document Format (pdf)) | ISBN 9783110557121 | ISBN 9783110558920 (e-book pdf) | ISBN 9783110557169 (e-book epub)
Subjects: LCSH: Impedance spectroscopy--Industrial applications. | Materials--Testing.
Classification: LCC TA165 (ebook) | LCC TA165 .I495 2018 (print) | DDC 535/.8--dc23
LC record available at https://lccn.loc.gov/2018030273

Bibliographic information published by the Deutsche Nationalbibliothek
The Deutsche Nationalbibliothek lists this publication in the Deutsche Nationalbibliografie; detailed bibliographic data are available on the Internet at http://dnb.dnb.de.

Cover image: Grotmarsel / iStock / Getty Images
Typesetting: VTeX UAB, Lithuania
Printing and binding: CPI books GmbH, Leck

www.degruyter.com

Preface

Impedance spectroscopy is a measurement method applied in many fields of science and technology, such as electrochemistry, material science, biology and medicine. The measurement method is based on a system identification approach and uses excitation signals of different frequencies to get more information about the system under test. It provides interesting possibilities to use the complex impedance of a material or a system over a wide frequency range. Thereby, effects dominating at different frequency ranges can be separated, non-accessible quantities become measurable, accurate measurements can be realized and multifunctional measurements can be carried out at low costs.

For Impedance Spectroscopy (IS), competency in several fields of science and technology is indispensable. An understanding of electrochemical and physical phenomena is necessary for modeling and understanding such phenomena. Measurement procedures should be developed taking specific requirements of the considered application into account. Signal processing methods are very important for extracting target information by suitable mathematical methods and algorithms. New trends are emerging rapidly, involving special techniques for realizing fully automatic embedded solutions at low costs and requiring a deep overview of modern information technology.

This book addresses impedance spectroscopy and its advanced applications in the fields of battery research, bioimpedance, measurement systems and sensors. It begins with a comprehensive review on impedance spectroscopy advances and future trends. Then, in three parts, it reports about new advances and different approaches in dealing with impedance spectroscopy, including theory, methodology and application aspects.

In the first part, the focus is on electrochemical and physical phenomena in the fields of nanocomposites, fuel, battery cells, nanoparticles and alloys. In the second part, measurement systems for impedance spectroscopy are discussed, including uncertainty, measurement duration, simulation, system implementation and interface circuits. In the third part, different sensors are presented, including capacitive sensors for detecting aqueous solutions, eddy current sensors for corrosion measurement of steel, humidity sensing and nitride sensing. In the fourth part, about bioimpedance spectroscopy, the topics are manifold and cover deep brain stimulation, choice of the number frequencies for measurement, intraventricular bioimpedance spectroscopy and meat quality monitoring by classification of impedance spectra.

The book scope is representative for the versatile measurement method with its highly interesting facets. In addition to recent advances concerning the main application fields of impedance spectroscopy, the book includes many methods and reflections on theory, modeling and implementation aspects. This book is therefore very interesting for researchers and developers in the field of impedance spectroscopy.

https://doi.org/10.1515/9783110558920-201

I thank all contributors for the interesting contributions and the reviewer who supported the development of this book. We hope together to contribute actively to further developments in this field by sharing our scientific and practical knowledge.

Prof. Dr.-Ing. Olfa Kanoun

Contents

Part II: **Measurement systems**

List of Contributing Authors

Paul Aurosmit
Chair for Measurement and Sensor Technology
Chemnitz University of Technology
Reichenhainer Strasse 70
09126 Chemnitz
Germany

Ammar Al-Hamry
Chair for Measurements and Sensor Technology
Chemnitz University of Technology
Reichenhainer Strasse 70
09126 Chemnitz
Germany
E-mail: ammar.al-hamry@s2011.tu-chemnitz.de

Anup Barai
WMG
University of Warwick
Coventry CV4 7AL
United Kingdom
E-mail: A.Barai@warwick.ac.uk

Thorsten Baumhöfer
Electrochemical Energy Conversion and Storage Systems Group
Institute for Power Electronics and Electrical Drives
RWTH Aachen University
Aachen
Germany
E-mail:
Thorsten.Baumhoefer@isea.rwth-aachen.de

Abderrahmane Benchirouf
Chair for Measurement and Sensor Technology
Chemnitz University of Technology
Reichenhainer Strasse 70
09126 Chemnitz
Germany

Serge Bernard
LIRMM – University of Montpellier
161 rue Ada
34095 Monptellier Cedex 5
France

Stephan Blokzyl
Chemnitz University of Technology
Faculty of Computer Science
Professorship of Computer Engineering
Chemnitz
Germany

Sylvain Bonhommeau
IFREMER
Rue Jean Bertho
97822 Le Port
France

Mariane Comte
LIRMM – University of Montpellier
161 rue Ada
34095 Monptellier Cedex 5
France

Nabil Derbel
National School of Engineers of Sfax
University of Sfax
Sfax
Tunisia

Renato Veiga de Torres
Chair for Measurements and Sensor Technology
Chemnitz University of Technology
Reichenhainer Strasse 70
09126 Chemnitz
Germany

Siham Yazami el Idrissi
Laboratory of Engineering Physics and Environment
Department of Physics, Faculty of Science
University Ibn Tofail
Kenitra
Morocco

Ahmed Fendri
Chair for Measurements and Sensor Technology
Chemnitz University of Technology
Reichenhainer Strasse 70
09126 Chemnitz
Germany
E-mail: ahmed.fendri@s2013.tu-chemnitz.de

Marju Ferenets
Eliko Competence Centre
Tallinn
Estonia

Hamadi Ghariani
National School of Engineers of Sfax
University of Sfax
Sfax
Tunisia

Rauno Gordon
Thomas Johann Seebeck Department of Electronics
Tallinn University of Technology
Tallinn
Estonia
E-mail: rauno.gordon@ttu.ee

Mahdi Guermazi
Chair for Measurement and Sensor Technology
Chemnitz University of Technology
Reichenhainerstrasse 70
09126 Chemnitz
Germany
and
National School of Engineers of Sfax
University of Sfax
Sfax
Tunisia
E-mail: mahdienis@yahoo.fr

My Mustapha Hafid
Laboratory of Engineering Physics and Environment
Department of Physics
Faculty of Science
University Ibn Tofail
Kenitra
Morocco

Wolfram Hardt
Chemnitz University of Technology
Faculty of Computer Science
Professorship of Computer Engineering
Chemnitz
Germany

Paul Jennings
WMG
University of Warwick
Coventry CV4 7AL
United Kingdom

Racem Jribi
National Engineering School of Gabes
Cabès
Tunisia

Olfa Kanoun
Chair for Measurement and Sensor Technology
Chemnitz University of Technology
Reichenhainer Strasse 70
09126 Chemnitz
Germany
E-mail: Olfa.Kanoun@etit.tu-chemnitz.de

Andres Käsper
Eliko Competence Centre
Tallinn
Estonia

Susanne Kersten
Fachgruppe Physik
exp. Teilchenphysik
Wuppertal
Germany

Vincent Kerzérho
LIRMM – University of Montpellier
161 rue Ada
34095 Monptellier Cedex 5
France

Yousra kessad
Laboratory of Engineering Physics and Environment
Department of Physics
Faculty of Science
University Ibn Tofail
Kenitra
Morocco

Andres Kink
SMARTIMPLANT Ltd.
Tallinn
Estonia
E-mail: andres@linking.ee

Hip Kõiv
Thomas Johann Seebeck Department of Electronics
Tallinn University of Technology
Tallinn
Estonia

Achraf Lamlih
LIRMM – University of Montpellier
161 rue Ada
34095 Monptellier Cedex 5
France
E-mail: Achraf.Lamlih@lirmm.fr

Raul Land
Thomas Johann Seebeck Department of Electronics
Tallinn University of Technology
Tallinn
Estonia

Noureddine Leguedani
Laboratory of Engineering Physics and Environment
Department of Physics
Faculty of Science
University Ibn Tofail
Kenitra
Morocco
E-mail: leguedani@yahoo.fr

Lukasz Macioszek
Institute of Metrology, Electronics and Computer Science
Faculty of Computer, Electrical and Control Engineering
University of Zielona Gora
Zielona
Poland
E-mail: l.macioszek@imei.uz.zgora.pl

Olev Märtens
Thomas Johann Seebeck Department of Electronics
Tallinn University of Technology
Tallinn
Estonia
E-mail: olev.martens@ttu.ee

Andrew McGorden
WMG
University of Warwick
Coventry CV4 7AL
United Kingdom

Mart Min
Thomas Johann Seebeck Department of Electronics
Tallinn University of Technology
Tallinn
Estonia

Christian Müller
Chair for Measurement and Sensor Technology
Chemnitz University of Technology
Reichenhainer Strasse 70
09126 Chemnitz
Germany

Jaan Ojarand
Thomas Johann Seebeck Department of Electronics
Tallinn University of Technology
Tallinn
Estonia
E-mail: Jaan.Ojarand@ttu.ee

Thomas Ortlepp
CiS Forschungsinstitut für Mikrosensorik GmbH
Erfurt
Germany

Michel Renovell
LIRMM – University of Montpellier
161 rue Ada
34095 Monptellier Cedex 5
France

Marek Rist
Thomas Johann Seebeck Department of Electronics
Tallinn University of Technology
Tallinn
Estonia
and
Eliko Competence Centre
Tallinn
Estonia

Susanne Rothgang
Electrochemical Energy Conversion and Storage Systems Group
Institute for Power Electronics and Electrical Drives
RWTH Aachen University
Aachen
Germany

Tristan Rouyer
IFREMER
Avenue Jean Monnet
34203 Sète
France

Ryszard Rybski
Institute of Metrology, Electronics and Computer Science
Faculty of Computer, Electrical and Control Engineering
University of Zielona Gora
Zielona
Poland

Abdulkadir Sanli
Chair for Measurement and Sensor Technology
Chemnitz University of Technology
Reichenhainer Strasse 70
09126 Chemnitz
Germany
E-mail: abdulkadir.sanli@s2012.tu-chemnitz.de

Dirk Uwe Sauer
Electrochemical Energy Conversion and Storage Systems Group
Institute for Power Electronics and Electrical Drives
RWTH Aachen University
Aachen
Germany

René Schmidt
Chemnitz University of Technology
Faculty of Computer Science
Professorship of Computer Engineering
Chemnitz
Germany
E-mail: rene.schmidt@informatik.tu-chemnitz.de

Fabien Soulier
LIRMM – University of Montpellier
161 rue Ada
34095 Monptellier Cedex 5
France

Jaqueline Stauffenberg
CiS Forschungsinstitut für Mikrosensorik GmbH
Erfurt
Germany

Markus Tahedl
Oertli Instrumente AG
Berneck
Switzerland

Ingo Tobehn-Steinhäuser
CiS Forschungsinstitut für Mikrosensorik GmbH
Erfurt
Germany
E-mail: itobehn@cismst.de

Christian Weber
ifm electronic gmbh
Essen
Germany
E-mail: christian.weber@ifm.com

Frank Wendler
Chair for Measurement and Sensor Technology
Chemnitz University of Technology
Reichenhainer Strasse 70
09126 Chemnitz
Germany
E-mail: frank.wendler@etit.tu-chemnitz.de

Heike Wünscher
CiS Forschungsinstitut für Mikrosensorik GmbH
Erfurt
Germany

Christian Zeitnitz
Fachgruppe Physik
exp. Teilchenphysik
Wuppertal
Germany

Olfa Kanoun

Impedance spectroscopy advances and future trends: A comprehensive review

Abstract: Impedance spectroscopy is a highly performant measurement method, which is often used in many fields of science and technology. It refers to the measurement of the complex impedance over a range of frequencies and provides possibilities for separating relaxation effects, accurate measurements and measurements of non-accessible quantities. In material science, impedance spectroscopy allows the characterization of the complex conductivity. Impedance spectra of battery cells allow to get information in a non-destructive way and to separate different phenomena concerning electrolyte and electrodes.In the biomedical field, it has the decisive advantage by providing a lot of information non-invasively and for long time intervals. For sensors, decisive improvements of measurement accuracy can be reached and multifunctional sensors can be realized, which allow measurements of more than one measurement quantity at the same time and same state of a material or a device.

In this contribution, after an introduction about the fundamentals of the measurement method, the links and differences between a wide range of application areas of impedance spectroscopy are shown for the first time. In each of these fields, we highlight the specific strengths of impedance spectroscopy and report on recent advances and future trends.

Keywords: Impedance spectroscopy, electrochemical impedance spectroscopy, relaxations, battery diagnosis, bioimpedance, sensors, biosensors, inductive sensors, capacitive sensors, nanocomposite, corrosion, signal processing

1 Introduction

The complex impedance of a material or a system provides information about its composition, interfaces and geometry and helps to follow changes of its state. Impedance measurement has therefore decisive benefits compared to other characterization methods, as it is experimentally efficient, because of the relative reduced time, hardware and software needed. It delivers more information than only resistive, capacitive or inductive measurements, because it considers at the same time both the real and the imaginary part of the complex impedance. It provides the possibility to separate

Acknowledgement: The author would like to thank the team of the professorship for measurement and sensor technology at the TU Chemnitz in the years 2007-2018 and the sponsors who have supported this work all over the years.

Olfa Kanoun, Chair for Measurement and Sensor Technology, Chemnitz University of Technology, Reichenhainer Strasse 70, 09126 Chemnitz, Germany, e-mail: Olfa.Kanoun@etit.tu-chemnitz.de

https://doi.org/10.1515/9783110558920-001

effects having linearly independent dispersions and dominating in different frequency ranges.

Impedance spectroscopy has shown a high potential for many years and has so far demonstrated its potential in a big number of applications, such as characterization of energy storage units [1], biomedical and biological applications [2] , [3] , [4], material testing [5–7], corrosion and coatings [8] and inductive [9, 10], capacitive [11] and conductive sensors [12]. The range of applications is very broad and the advantages it provides are significant under the condition that it is applied in the right manner. In every application, field-specific strengths of the method are in the focus of interest, even if the method can provide more advantages.

Impedance spectroscopy is non-invasive, as it uses small amplitudes of the excitation signal. It is capable to achieve real-time measurement by using wide frequency band signals. Even if it is generally known as a method for materials and systems characterizations in laboratory environments, impedance spectroscopy is increasingly used for real-time measurements in portable and embedded solutions due to the favorable developments in the field of microelectronics.

In this contribution, we give an overview about different aspects of the method in relation to a wide range of application fields. The aim is to demonstrate the potential of impedance spectroscopy and to show the link between the different application areas. In each context, the reasons for the particular strengths of the method are discussed and special potentials are shown. Then, recent developments are explained by means of selected examples.

2 Theoretical background

Impedance spectroscopy belongs to the so-called system identification techniques, where the frequency-dependent transfer function of a system is determined. In impedance spectroscopy, the complex impedance of a material or a system is measured at different frequencies or in a certain frequency range. Thereby both the real and the imaginary part of the complex impedance can be considered as useful sources of information, where certain phenomena are dominating in certain frequency ranges and can therefore be characterized. In general, the real and the imaginary part of the impedance show a different frequency dependency and therefore provide a comprehensive source of information. This is why manifold information regarding electrical and electrochemical properties of the investigated system or material can be extracted.

A wide frequency band helps to gain more information at a given working point of the i-u characteristic (see Fig. 1 (a) and (b)). If the measurements carried out correspond to the same phenomena (see Fig. 1 (c)), impedance spectroscopy helps to realize a better characterization of the phenomena. The measurements are from a physical point of view redundant and provide possibilities to have a better measurement

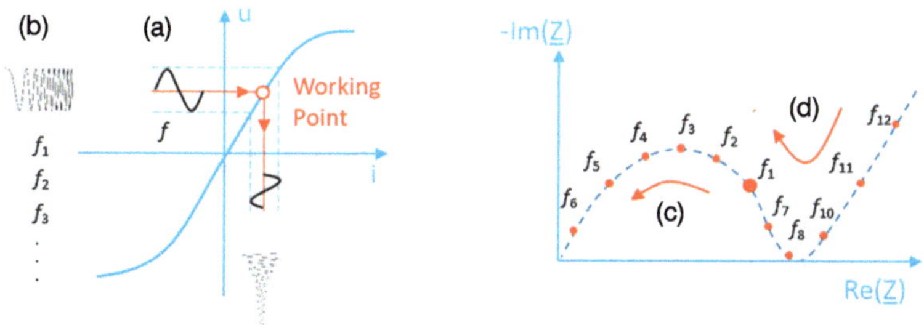

Fig. 1: Principle of impedance spectroscopy. (a) Excitation with a single frequency at a given working point. (b) Example for excitation with a varied frequency with a sweep. (c) Improvement of measurement information by redundancy. (d) Improvement of measurement information by diversity trough a wide band excitation signal.

accuracy, as it becomes possible to reduce the influence of random deviation, such as noise, and systematic deviations, e. g., temperature, and to benefit from a detailed modeling of the characteristic.

If the impedance measurement is carried out in a wider frequency range (see Fig. 1 (d)), the information gained thereby becomes more probable to have diversity, as different physical/chemical phenomena can be dominating in different frequency ranges. If different spectral behaviors are included in the impedance spectrum referring to different phenomena, impedance spectroscopy provides interesting opportunities to measure more than one quantity at the same time. The conditions for this are (i) to have sufficient measurement data in the spectral regions where the individual effects are dominating and (ii) for the time constants corresponding to these phenomena to be sufficiently distant.

3 Extraction of information from impedance spectra

Several aspects should be specifically addressed on the design process for an impedance measurement system, such as measurement procedures, investigations of physical and chemical phenomena taking place, development of suitable impedance models and extraction of target information by optimization techniques.

In general, there are different approaches, which can be considered for information extraction from impedance spectra. Physical methods are interesting and allow profound insights in a material or a system (see Fig. 2). Mathematical methods have also been proposed, such as data mining, principal component analysis [13], neural networks and fuzzy logic [14].

For physical modeling, it is necessary to understand the physical and chemical phenomena taking place within the system and to formulate the necessary model equations by combining the descriptions of partial phenomena, e. g., in series, par-

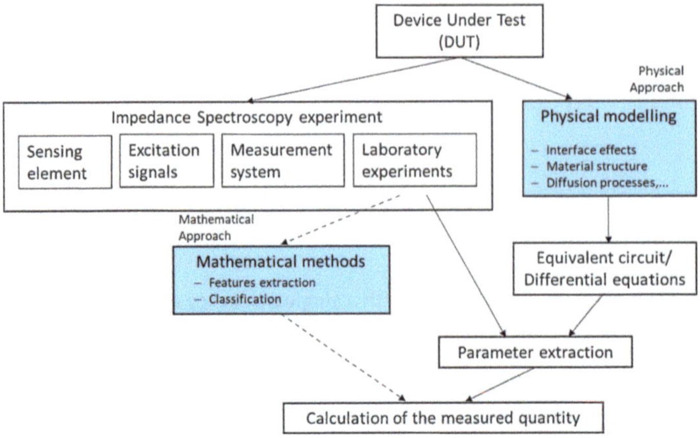

Fig. 2: General approach of impedance spectroscopy and for information extraction from impedance spectra.

allel or cascade. Thereby, different formulations can be mathematically equivalent, but only some of them provide a meaningful model and are therefore physically relevant. The extraction of information is carried out, in this case, by determination of the model parameters, which are directly related to the corresponding physical phenomena. If the formulation of the model is suitable and the parameter extraction is successful [15], the model parameter provides information about the measured quantity and shows a physically expected behavior. In this case, extracted parameters inferring the measurement quantity are directly used for the calculation of the measurement quantity.

Mathematical methods are also possible, but need a big data base with relevant experiments in order to build the necessary basis for signal processing and extraction of information. Suitable feature extraction and classification methods need to be developed in this case to discriminate between the different classes corresponding to the state of the system [14]. For this purpose, the training data and the experimental data need to be representative for the real scenario and include all relevant cases that could influence the results, such as environmental parameters, aging and contamination effects.

4 Impedance spectroscopy in measurements and sensor solutions

Impedance spectroscopy provides the possibility to follow the state of a system and to track its changes, e. g., due to aging or chemical reactions taking place. This is the case, e. g., for corrosion, battery diagnosis and bioimpedance. Especially for battery diagnosis and bioimpedance analysis, impedance spectroscopy allows to non-invasively get information about the state of a cell or a battery and to determine specific information concerning parts of it or phenomena taking place within it. In the case of conductive, capacitive and inductive sensors, impedance spectroscopy allows to measure quantities in spite of experimental limitations, to improve the accuracy or even to measure more than one quantity at the same time.

The applications for impedance spectroscopy are therefore manifold. In the next sections, we present a selection of the most important applications for impedance spectroscopy (see Fig. 3) and show the potential of this measurement method in the corresponding fields in comparison with other classical methods.

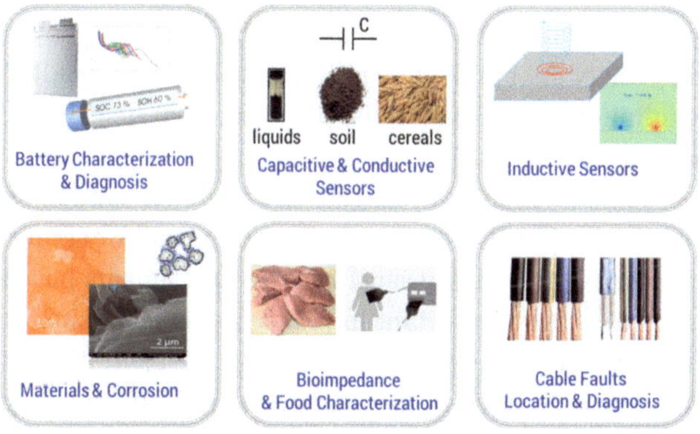

Fig. 3: Selection of impedance spectroscopy applications in measurement and sensor solutions.

5 Material characterization

For materials, impedance spectroscopy provides interesting possibilities to characterize the complex electrical behavior including resistive, capacitive and inductive behavior [5]. Thereby, it is sensitive to both surface phenomena and changes of bulk properties including conduction mechanisms.

For example, impedance spectroscopy is a well-established method, which allows quantitative characterization of corrosion processes as well as the evaluation of the anti-corrosion performance of protective coatings on corrosive metals [16]. It can be used for several purposes, among which the detection of surface and material changes due to exposure, prediction of the lifetime of corrosion protection, evaluation of coating systems, identification of the corrosion processes that lead to failures, measurement of water uptake by coatings and the development of models for coating/metal system performance.

One of the important advantages of this method is that it is a non-destructive technique. It can follow the evolution of coated metals as they undergo either natural or artificial exposure to conditions that cause corrosive failure to provide information about the corrosion kinetics. Impedance measurements allow the prediction of the long-term performance of the coatings and track very effectively the effects of temperature, water ingress, coating plasticization and the glass transition temperature in coatings [17].

Impedance spectroscopy is interesting for electrical characterization of almost every material. It has played a big role for many years in the field of nanocomposites, which are becoming increasingly important in the field of lightweight structures and sensors. Flexible films of carbon nanotubes (CNTs) can be prepared for several sensor applications, such as strain gauges or pressure sensors showing a high sensitivity and performing under low costs. With less than 1 % filler material, input conductive and percolation-based CNT polymer composite sensors can be fabricated. Thereby, a suitable composition and homogeneity are highly demanded and should be fulfilled by the fabrication process including choice of materials, preparation of dispersion, substrate treatment and deposition techniques.

Impedance spectroscopy enables to deeply investigate the complex conductivity of the nanocomposite materials, so that they can be better optimized for certain required properties [5, 18]. For example, for highly sensitive pressure sensors based on CNT polymer nanocomposites, it is very important to realize nanocomposites at the percolation ratio [19]. The percolation ratio is the nanofiller concentration at which rarely conductive paths can be built through the nanofillers without tunneling through the polymer. At this concentration, the nanocomposite becomes more sensitive to pressure, because an increase of pressure leads to a reduction of the distance between the CNTs and therefore increasingly leads to percolation. These concentrations cannot be determined easily by experimental methods due to fluctuating processing parameters during the fabrication process.

Impedance spectroscopy is suitable for characterization of the percolation behavior, as it includes the information about both resistive and capacitive behaviors of the nanocomposite [5]. Especially the imaginary part of the impedance gives information about the average gap between nanofillers and is therefore very important for investigating percolation (see Fig. 4).

In [5] it has been shown that impedance spectroscopy measurements allow the experimental determination of the percolation threshold of the nanocomposite. At concentrations lower than the percolation ratio, the nanocomposite shows an almost purely capacitive behavior. Due to inter-tube distances, the share of capacitive behavior dominates the conduction behavior (see Fig. 4 (a)). Beyond the percolation ratio, the conductive paths dominate (see Fig. 4 (b)) and show a frequency independent behavior. The percolation region is exactly the region where the transition between these two extreme cases takes place abruptly.

This example shows that, by means of impedance spectroscopy, the percolation ratio can be determined experimentally even after film deposition. This allows tailor-

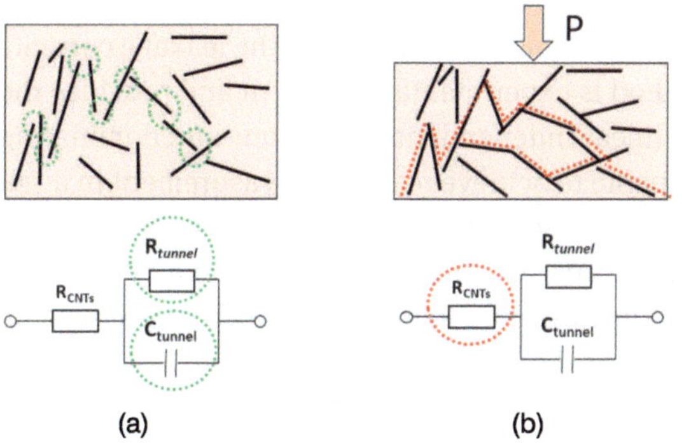

Fig. 4: CNT polymer composite. (a) Relaxed: tunneling effect, the capacitive behavior is dominating. (b) Under pressure: resistive behavior is dominating due to increasing contacts between CNTs.

ing of the piezoresistivity of strain-sensitive films based on nanocomposites for different nanofillers and even by fluctuating processing parameters.

6 Battery characterization and diagnostics

Modern battery management systems need accurate information concerning the state of charge (SoC) [20], the state of health (SoH) [21], the battery lifetime, the internal temperature and fault diagnosis within cell packs. Other methods are, e. g., voltage measurement and coulomb counting, which bring some challenges in the case of Li-ion batteries. In particular, the SoC-vs-voltage curve is very flat, so that the sensitivity of the voltage to the SoC is very low. The direct voltage measurement is not sufficient for a comprehensive diagnosis as it is dependent on SoC, temperature and current level at the same time. In general, the open-circuit voltage (OCV) provides an alternative, which can be determined at offline state, where the battery is isolated from the external circuit for a long time. In online measurement, it can be estimated from the voltage measurement by considering the internal resistance multiplied by the current [22] Coulomb counting integrates the current over time to estimate the remaining battery charge. But, due to the limited sampling rate, several current impulses may not be tracked with a high accuracy. Therefore, the coulomb counting drifts over time due to the cumulative deviation.

In contrast, impedance spectroscopy is a non-destructive method, which can estimate the remaining charge of the battery, also known as SoC and the state of health (SoH) of the battery. This is due to the fact, that battery impedance increases with aging and changes due to temperature and also different phenomena taking place at low, middle and high frequencies. Therefore, impedance spectroscopy can be used as a diagnostic tool and accurate analytical technique for measuring different critical battery parameters. It provides the possibility to get information about the inner state of the battery, including both the electrolyte and the electrodes, non-invasively and in a relatively short measurement time, without needing to carry out any cycling processes.

Impedance spectroscopy offers a solution that is independent on the leakage currents and the fluctuating voltage. This method is experimentally efficient and provides the possibility to separate effects having linear independent dispersions and dominating in different frequency ranges. It is capable to achieve real-time measurement in a relatively short measurement time as it can use wide frequency band excitation signals. As a battery ages, several phenomena occur, accompanied by a change of the overall impedance of the battery [22–24]. The SoC of the battery can be also estimated by means of impedance spectroscopy [25–28].

The battery impedance spectrum includes information about the electrolytes, diffusion layer and electrode reactions, as explained in Fig. 5. At low frequencies, in the range of mHz, processes with a high time constant are stimulated, such as migration and diffusion. Diffusion effects are dominating in this frequency range. At middle frequencies, electrode reactions dominate and may show a constant phase behavior due to electrode porosity. At high frequencies, inductive effects in the battery wiring and porous structure are dominant.

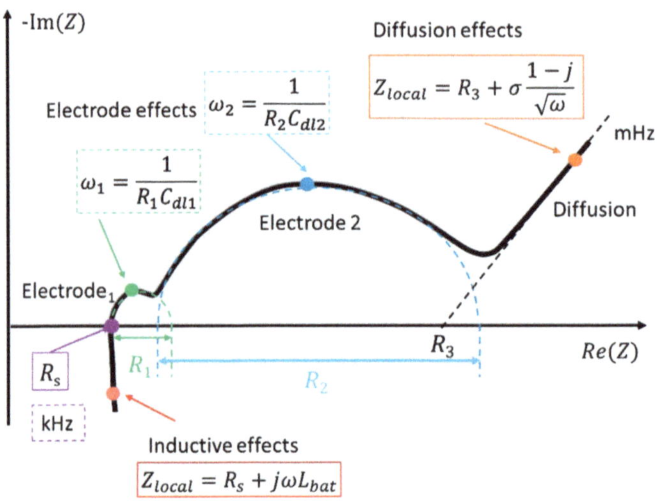

Fig. 5: Typical impedance spectrum of a battery in an exaggerated Nyquist plot. Here, R_s is the series resistance, R_1 and R_2 are the charge transfer resistances of the electrodes, C_{dl1} and C_{dl2} are the double layer capacitances of the electrodes, σ is the Warburg coefficient, R_3 is the intersection of the diffusion characteristic with the real impedance axis and L represents inductive effects.

The individual phenomena in the impedance spectrum of a cell have their own models, which can be expressed analytically or in form of equivalent circuits. Models based upon concentrated circuit elements such as capacitors, resistors and inductors can simulate some internal electrical properties of the cell. Nevertheless, especially for diffusion processes and due to electrode porosity, elements with distributed parameters, such as constant phase elements and Warburg impedance, are also necessary to model the impedance behavior accurately.

The impedance spectrum of a battery cell includes manifold information and the method is therefore very interesting for diagnosis. Nevertheless applying it on a battery cell involves many challenges. The low-frequency measurements at mHz frequencies need a long measurement time and involve therefore a risk of charging or discharging the battery cell and changing its state thereby. The battery impedance is low and needs to be measured with a high resolution. A big challenge in modeling impedance spectra is that the individual phenomena may overlap in the impedance spectrum. This overlapping is strongly dependent on the cell technology and also on the cell state. In the case of overlap, the separation of effects and the parameter calculation become more difficult, especially if the measured frequency is limited. The AC current distribution within cylindrical Li-ion cells depends strongly on the frequency and the temperature [29]. This may lead to limitations for deep interpretation. The cell design may overshadow some active material impedance by local inhomogeneity due to lithium plating or advanced aging. The spectrum may change its structure due to aging and models need to be adaptively adjusted to consider this effect.

Yet, the latest developments allow to realize embedded solutions for impedance spectroscopy, which realize a comparable measurement quality like sophisticated laboratory devices [21, 30] having low cost and a significantly short measurement time. This development allows to use impedance for real-time measurements on site and to determine useful information for intelligent battery management systems, such as SoH [22], SoC [20], the internal cell temperature [31] and the remaining useful battery life [32].

7 Biomedical applications and food characterization

For non-invasive tissue characterization, pathological health status investigation and diseases diagnosis, bioimpedance spectroscopy is one of the most important methods. In comparison with classical non-invasive techniques used in clinical research, such as optical spectroscopy, ultrasound and X-ray, bioimpedance spectroscopy is easy to use and safe. It can give information on the deepness of a material and helps to avoid interferences with substances reacting in the same wavelength. Therefore, impedance spectroscopy is gaining importance in the biological and food sectors and will play a more important role for the food industry in the future.

Impedance spectroscopy can be used for characterization of different biological tissues and for medical [33–37] and food applications [11, 14, 38–41], because the complex impedance gives information about the composition and the state of a biological tissue or organ. Although there are many different methods, impedance spectroscopy is often preferred.

The impedance of a muscle includes three areas of dispersion: the α dispersion at low frequencies between some Hz and some kHz, β dispersion at middle frequencies in the range from some kHz to MHz and γ dispersion at higher frequencies [14, 33, 38, 39]. The β dispersion is very informative about the state of the muscle as it provides information about the cell membranes. Therefore, impedance spectroscopy is a very promising method for assessment of meat freshness and meet characterization in general.

A lot of challenges should be overcome in the measurement of muscle impedance in order to guarantee reproducible measurements. For this, an electrode configuration has been developed to avoid influence of anisotropy and in order to realize a good contact to the biological material in spite of its fibrous structure [4, 42]. The developed probe consists of one central cylindrical electrode and eight surrounding needle electrodes made of gold-plated steel (ϕ of 1 mm) in order to avoid corrosion.

The β dispersion, corresponding to radio frequency, is strongly dependent on the dielectric properties of the cell membranes and their interactions with the extra- and intracellular electrolytes, as the current passes through the cell membranes at these frequencies. It includes information referring to the behavior of the cell membrane integrity, which changes significantly with aging (s. Fig. 6). One of the phenomena that contributes to the β dispersion is the Maxwell–Wagner effect related to the interfacial relaxation process occurring in systems where the electric current passes at the interface between two different materials.

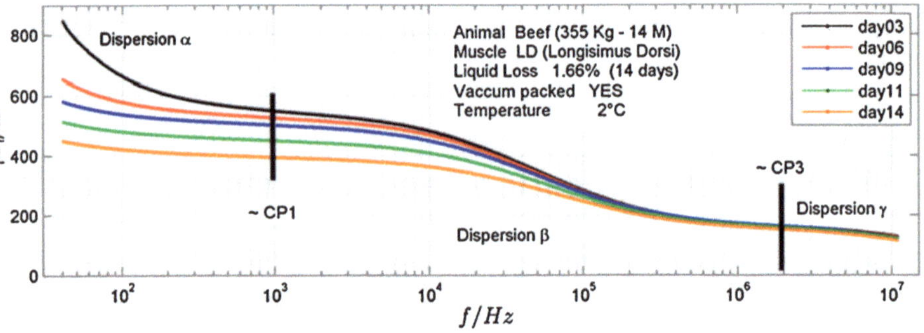

Fig. 6: Bioimpedance measurement of a beef longissimus dorsi muscle for 14 days [14].

In [43], a method for measuring bioimpedance non-invasively is presented. It consists of magneto-inductive measurement techniques, such as gradiometers. Gradiometers consist of a sending coil and two symmetrical receiving coils. For the detection of tissue impedance, they use the difference principle between the coil on the tissue side and the coil on the other side. They can therefore detect even small changes of electrical properties of a biological tissue. This high sensitivity has the drawback that they may have different levels of detuning, which cannot be easily adjusted. In [43], a concept for adjusting the gradiometer in a predefined initial state is presented. It uses additional coils on both sides to adjust the detuning by a suitable current flow through the additional coils and realize an accurate measurement.

Measuring bioimpedance of living beings in medical applications brings several challenges. One is the limitation of the current, which must be applied without doing any damage [46]. The medical requirements imposed by the standard IEC-60601 claim that the current should be below the perception threshold of 0.5 mA at 5 kHz [44]. The perception threshold increases when the frequency increases following the equation [45]

$$I_{\text{max}} = 10^7 f, \tag{1}$$

where the current is the root mean square value and the frequency is in Hz.

Therefore, for *in vivo* applications, a current excitation is preferred to a voltage excitation, which can be affected by contact impedance between the electrode and the tissue and high currents through the tissue due to impedance changes or temperature dependence of the tissue conductivity.

One further challenge is the fact that living beings change their state, e. g., due to breathing, moving and blood circulation. This is why one of the approaches which can be considered is to reduce the measurement time in order to maintain the system under test under quasi-static conditions. Many researchers investigated possibilities for using wide band signals including different frequencies in order to get a spectrum in a reduced measurement time [35, 47–49]. These are typically, e. g., chirp signals [35], multisine signals, binary chirp and binary multifrequency signals [49].

In [50], a novel measurement and identification approach, called the parametric-in-time approach, has been developed for time-varying bioimpedance systems with a quasi-periodic character. This allows to conduct myocardial impedance measurements *in vivo* and *in situ*.

8 Cable fault detection and localization

The increasing use of the wiring in vehicles, production plants, communication and power systems leads to a big nead for wiring faults detection and location. Experience in this area indicates that the majority of the problems encountered are due to cable errors and increased contact resistance of connectors. If the cables are long, not all methods are able to detect the fault position as signals may become weak and even concealed by noise.

There are several methods for wire fault location and characterization, of which time domain reflectometry (TDR) is the most common one. In this case, a pulse signal is sent on the cable, which is supposed to be homogenous. If the cable has a fault, a signal is reflected and this reflection can be measured to detect the position of the fault by estimating the time of flight and the type of the reflection includes information about the type of change of the cable impedance and therefore about the type of the fault.

Over the last decade, many methods, such as frequency domain reflectometry (FDR), ultra-wide band-based TDR, spectrum TDR (STDR) and impedance spectroscopy (IS), have been developed. They use different wide band incident signals and signal processing methods, but the aim was mainly to improve the accuracy of the fault location.

In [6], impedance spectroscopy was introduced to identify both position and type of cable errors. The innovation in this approach is that the load impedance of the wire fault can be quantitatively extracted from its input impedance, so that the types of the wire faults can be exactly identified. For that, a transformation to the pseudo-frequency domain is applied to localize the wire fault, which corresponds to the periodicity of the spectrum represented in Fig. 7.

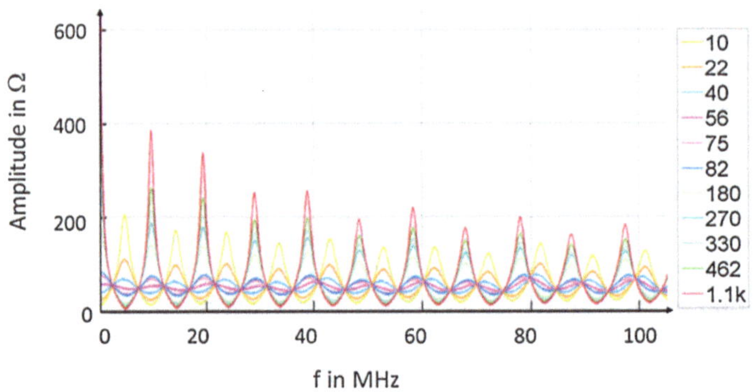

Fig. 7: Impedance spectrum of a 10-m coaxial cable with different terminating resistors. The amplitude is influenced by the terminating resistors and the periodicity of the impedance is influenced by the error position [6].

The input impedance of the wire at low frequencies is applied to identify the types of wire faults. The imaginary part of the cable impedance can be applied to detect the type of the wire fault. In the case of an open circuit, it is the capacitance of the cable. In the case of a short circuit, the inductance of the cable helps to detect the fault type.

9 Conductive, capacitive and inductive sensors

Impedance measurement has decisive benefits, as it is experimentally efficient and can be realized in low-cost embedded systems. Its advantages are particularly important in sensor technology as it delivers more information than only resistive, capacitive or inductive measurements. This is due to the fact that it uses both the real and the imaginary part of the complex impedance at different frequencies. It provides the possibility to separate effects having linearly independent dispersions and dominating in different frequency ranges.

Impedance-based sensors can have a variety of designs and can serve for the acquisition of various measured variables (see Fig. 8 and Fig. 9). These include, e. g., position, humidity, material properties and filling level. Very often, only the real part is measured in the case of conductive sensors and the imaginary part in the case of inductive and capacitive sensors. The other part of the impedance is considered as losses and is in general ignored, even if it includes also information about the "device under test". In some cases, the amplitude of the impedance and the measured variable is determined at a suitable sensitive frequency [53].

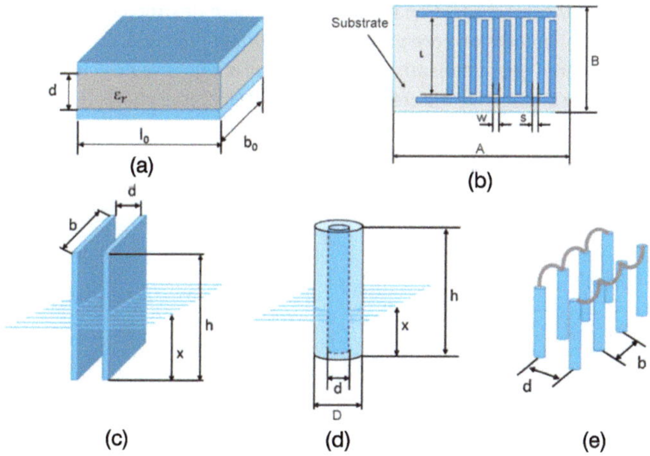

Fig. 8: Electrode structures for capacitive and conductive sensors.

The measurement of the impedance with its real and imaginary part at different frequencies expands the information considerably [54]. The information obtained from the sensor is representative of the measurements of a sensor array, in which each sensor is fictitiously operated at a frequency. We know from the field of multisensors that this increases the quality and reliability of the measurement and helps to achieve a better accuracy [55].

Impedance spectroscopy provides interesting possibilities for realizing a multifunctional measurement, so that one sensor measures several quantities at the same time. This provides interesting possibilities not only to reduce costs, but also to extend the information diversity from one sensor. In [13], an impedimetric sensor has been developed for the measurement of soil moisture, having a design such as in Fig. 8 (c). It measures both the real and the imaginary part of the impedance at different frequencies and is thereby able to identify the type of the soil. Similar sensors measuring only

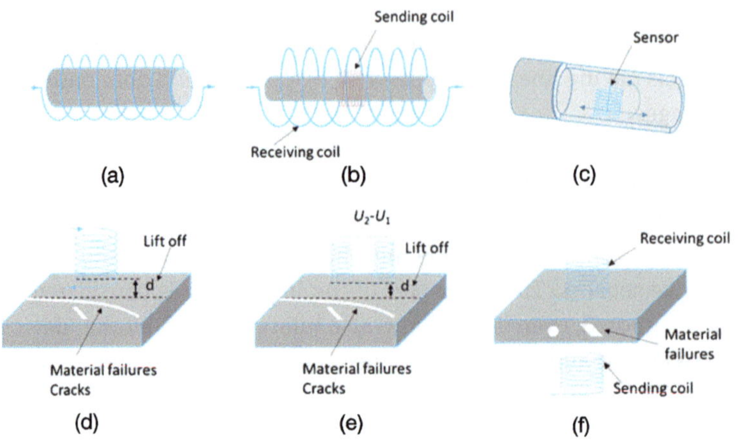

Fig. 9: Examples of inductive sensors for characterization of materials and non-destructive material testing. (a) immersion core principle, (b) Immersion core principle transformer principle, (c) inner wall characterization of a pipe, (d) material characterisation and testing by a reflective principle, (e) material characterisation and testing by a reflective difference principle, (f) material characterisation and testing by a transmissive principle.

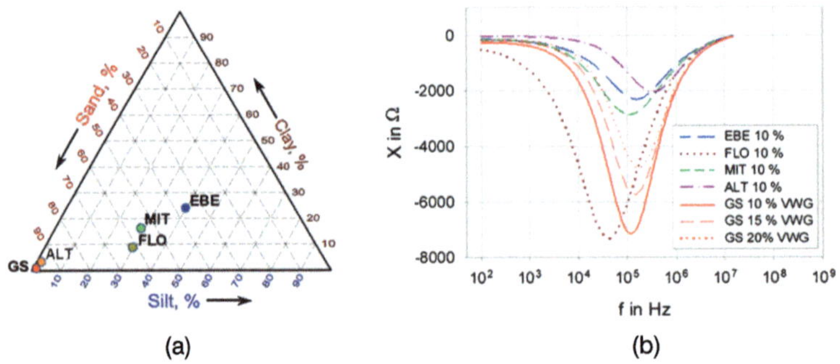

Fig. 10: Dependency of the imaginary part of the soil impedance on soil type and soil moisture. (a) Investigated soil types classified by their inorganic components [13]. (b) Soil imaginary part in dependence of the frequency for different soil types.

the capacitive part have massive problems with soil type influence, which leads to a high inaccuracy and needs therefore to be regularly calibrated to the corresponding soil. As can be seen in Fig. 10 (b), the changes of the imaginary part of the impedaance due to moisture show a different behaviour depending on the soil type. Measuring the capacity at a certain frequency, leads to ambiguityan is not sufficient for realizing an accurate measurement. In the case of the impedimetric sensor, the measurement data serve first to the classification of the soil type by means of data reduction via principal components analysis and classification via the k-nearest-neighbor method. After soil type recognition, the moisture measurements become easy and accurate by using a linear characteristic of the imaginary part of the admittance [13].

Inductive sensors and especially eddy current sensors can be used in different manners for both distance measurement and material characterizations. The main advantages of these sensors are a.o. their robustness [56], their process capability and the fact that they operate contactless and in a non-destructive manner [57].

Yet, if the imaginary part changes more, also the real part includes interesting information. A change of frequency leads to a different penetration depth of the magnetic field in the target material (see Fig. 11) and the coil impedance can therefore give information about cracks and failures under the surface.

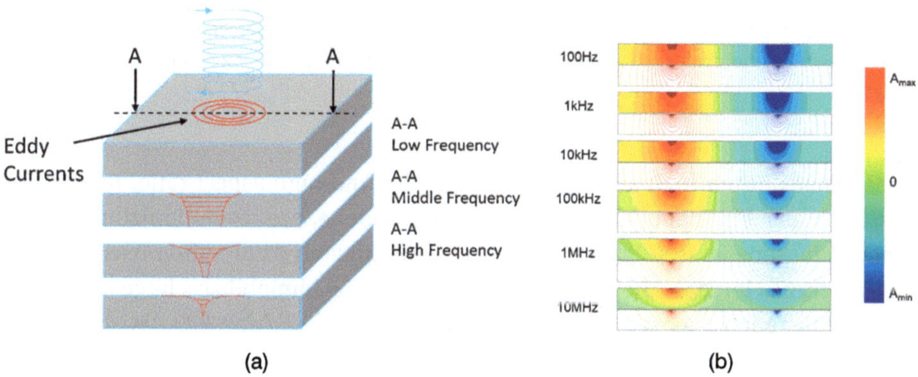

Fig. 11: Dependence of the eddy current effect on the frequency. (a) Schematic description of the penetration of a magnetic field in a target material in dependence of the frequency. (b) Numerical simulation of the magnetic vector potential in a ferric target material in dependence of the frequency.

10 Biosensors

Impedance spectroscopy has recently gained a high popularity for biosensing of a wide range of substances owing to its label-free measurements. It reaches thereby a superb sensitivity down to the femto- and attomolar ranges [58–60]. It allows rapid screening of biocompatible surfaces and monitoring pathogenic bacteria, as well as the analysis of biological cells and tissues.

The impedance spectroscopy transducer signal is in this case mainly based on the interactions between a certain biological receptor and the target species, which is selectively adsorbed from the solution (see Fig. 12) [61, 62]. Such interactions cause a change of the interfacial electron transfer kinetics between the solution species and the conducting electrode. The charge transfer resistance increases therefore with the increase in the quantity of targets bound to the receptive surface. Thereby, the selectivity, sensitivity, response time and detection limit are strongly dependent on the use of proper materials and catalysts [61, 63].

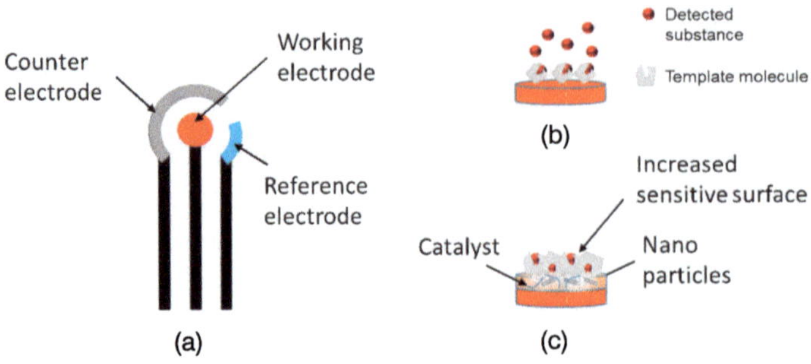

Fig. 12: Three-electrode electrochemical sensor. (a) Sensor element. (b) Selectivity enhancement by using template molecules. (c) Strategies for improving sensitivity by using nanoparticles and catalysts.

Because of their high affinity to biomolecules and high aspect ratio, nanomaterials, such as metal nanoparticles, CNTs and graphene, are revolutionizing electrochemical biosensors (see Fig. 12 (c)) [64, 65]. Nanomaterials are often used as electrode coating to increase their effective surface area and immobilize a large number of biomolecules such as enzymes and antibodies [65].

11 Future trends of impedance spectroscopy

The impedance spectroscopy is versatile as a method and its importance is in general increasing, as it provides interesting possibilities for characterizing materials, systems and sensors. Sensors themselves today tend to the use of high-speed and low-cost electronic circuits, allowing for increasingly more signal processing and advanced manufacturing technologies [52]. Similar aspects can be observed nowadays for impedance spectroscopy. Measurement devices use novel technologies and provide increasingly efficient solutions, enabling the use of the method in a wider range of applications.

11.1 Embedded and miniaturized solutions

Impedance spectroscopy was regarded as an electrochemical laboratory method for several decades. Today, due to the excellent developments in microelectronics, a cost-effective electronic realization becomes more and more possible. The most available embedded solutions are in general in the medical field and a lot of them are in medical implants. This trend is now increasing towards manifold solutions for sensors and diagnostic devices. In most cases, the aim is not to operate a large measuring range, but rather to find solutions that are usually limited by the requirements of a particular application.

11.2 Fast impedance spectroscopy and tomography

One of the important conditions for impedance spectroscopy is that the system does not change during measurement of the whole impedance spectrum. If the system has a higher dynamic, impedance measurement should be carried out within a short time to realize quasi-stable conditions. This is why reducing the measurement time is necessary to widen the scope of applications of impedance spectroscopy to more dynamic applications.

Many examples can be seen in impedance tomography. In this case, measurements with different electrode combinations contribute to the reconstruction of one image across the device under test. All these measurements need to be adjusted to the same system state. In order to fulfil this condition, in general only one or two relative high frequencies are measured. The development of fast impedance spectroscopy methods would enable tomography methods to better use the spectral information by involving more measurement frequencies.

11.3 Impedimetric sensors

Nowadays, in the sensor field there is an interesting trend towards multifunctional sensors, which are able to measure more than one quantity at the same time [51, 52]. Impedimetric sensors can provide more information as the impedance is measured and in a certain frequency range. The comprehensive data generated at one value of the measured quantity provide interesting possibilities to reduce influencing effects, improve the reliability of the information and even realize the measurement of more quantities at the same time. In this way, many classical capacitive, resistive or inductive sensors can reach a higher performance by using impedance spectroscopy.

Bibliography

[1] U. Tröltzsch, O. Kanoun, and H.-R. Tränkler, "Characterizing aging effects of lithium ion batteries by impedance spectroscopy", Electorchimica Acta, vol. 51, no. 8–9, pp. 1664–1672, 20 January 2006.

[2] D. Bouchaala, O. Kanoun, and N. Derbel, "Portable bioimpedance spectrometer for total frequency range of β dispersion", Technisches Messen, vol. 80, no. 11, pp. 373–378, November 2013.

[3] M. Min, T. Parve, V. Kukk, and A. Kuhlberg, "An implantable analyzer of bio-impedance dynamics: Mixed signal approach", IEEE Transactions on Instrumentation and Measurement, vol. 51, no. 4, pp. 674–678, August 2002.

[4] M. Guermazi, O. Kanoun, and N. Derbel, "Reduction of anisotropy influence and contacting effects in in-vitro bioimpedance measurements", Journal of Physics Conference Series, vol. 434, 12058, 2013.

[5] A. Sanli, C. Müller, O. Kanoun, C. Elibol, and M. F.-X. Wagner, "Piezoresistive characterization of multi-walled carbon nanotube-epoxy based flexible strain sensitive films by impedance spectroscopy", Composites Science and Technology, vol. 122, pp. 18–26, 2016.

[6] Q. Shi and O. Kanoun, "Wire fault location in coaxial cables by impedance spectroscopy", IEEE Sensors Journal, pp. 4465–4473, June 2013, doi: 10.1109/JSEN.2013.2269218.

[7] Q. Shi and O. Kanoun, "Model-based identification of wire network topology", Measurement, vol. 55, pp. 206–211, September 2014, Elsevier.

[8] H. Zarrok, A. Zarrouk, B. Hammouti, R. Salghi, C. Jama, and F. Bentiss, "Corrosion control of carbon steel in phosphoric acid by purpald – Weight loss, electrochemical and XPS studies", Corrosion Science, vol. 64, pp. 243–252, November 2012.

[9] F. Wendler, P. Büschel, and O. Kanoun, "Efficient parameter estimation impedance-based sensors by local linear transformation", Technisches Messen, vol. 81, no. 9, pp. 450–456, 2014.

[10] U. Tröltzsch and O. Kanoun, "Generalization of transmission line models for deriving the impedance of diffusion and porous media", Electrochimica Acta, vol. 75, no. 1, pp. 347–356, 2012.

[11] A. Fendri, H. Ghariani, and O. Kanoun, "Dielectric spectroscopy for assessment of water content in edible oils", In: 14th International Multi-Conference on Systems, Signals and Devices, March 28–31, 2017, Marrakech, Morocco, 2017.

[12] R. Gruden, A. Buchholz, and O. Kanoun, "Electrochemical analysis of water and suds by impedance spectroscopy and cyclic voltammetry", Journal of Sensors and Sensor Systems, vol. 3, pp. 133–140, 2014.

[13] A. Tetuyev and O. Kanoun, "Method of soil moisture measurement by impedance spectroscopy with soil type recognition for in-situ applications", Technisches Messen, vol. 73, no. 7/8, 2006.

[14] M. Guermazi, O. Kanoun, and N. Derbel, "Investigation of long time beef and veal meat behaviour by bio-impedance spectroscopy for meat monitoring", IEEE Sensors Journal, Sensors-9401, 2014.

[15] P. Büschel, U. Tröltzsch, and O. Kanoun, "Use of stochastic methods for robust parameter extraction from impedance spectra", Electrochimica Acta, vol. 56, 2011, Elsevier.

[16] G. Bierwagen, D. Tallman, J. Li, L. Hea, and C. Jeffcoate, "EIS studies of coated metals in accelerated exposure", Progress in Organic Coatings, vol. 46, pp. 148–157, 2003.

[17] M. Tezeghdenti, N. Etteyeb, L. Dhouibi, O. Kanoun, and A. Al-Hamri "Natural products as a source of environmentally friendly corrosion inhibitors of mild steel in dilute sulphuric acid: Experimental and computational studies", Protection of Metals and Physical Chemistry of Surfaces, vol. 53, no. 4, pp. 753–764, 2017.

[18] O. Kanoun, Chr. Müller, A. Benchirouf, A. Sanli, A. Bouhamed, A. Al-Hamry, and L. Bu, "Potential of flexible carbon nanotube films for high performance strain and pressure sensors", In: "Nanotechnology for Optics and Sensors", One Central Press, 2015, pp. 148–183.

[19] A. Bouhamed, Chr. Müller, S. Choura, and O. Kanoun, "Processing and characterization of MWCNTs/Epoxy nanocomposites thin films for strain sensing applications", Sensors and Actuators A, Physical, vol. 257, pp. 65–72, 15 April 2017.

[20] A. Cuadras and O. Kanoun, "SoC Li-ion battery monitoring with impedance spectroscopy", In: Sixth International IEEE Multi-Conference on Systems, Signals and Devices SSD'09, 23–26 March 2009, Jerba, Tunisia.

[21] A. Otto, S. Rzepka, T. Mager, B. Michel, C. Lanciotti, T. Günther, and O. Kanoun, "Battery management network for fully electrical vehicles featuring smart systems at cell and pack level", In: 16th International Forum on Advanced Microsystems for Automotive Applications, AMAA 2012, pp. 3–14, 30–31 May 2012, Berlin, Germany.

[22] M. Berecibar, I. Gandiaga, I. Villarreal, N. Omar, J. Van Mierlo, and P. Van Den Bossche, "Critical review of state of health estimation methods of Li-ion batteries for real applications", Renewable and Sustainable Energy Reviews, vol. 56, pp. 572–587, 2016.

[23] D. Stroe, M. Swierczynski, A. Stroe, and R. Teodorescu, "Lithium-ion battery power degradation modelling by electrochemical impedance spectroscopy", IET Renewable Power Generation, vol. 11, no. 9, pp. 1136–1141, 2017.

[24] U. Tröltzsch, O. Kanoun, and H.-R. Tränkler, "Characterizing aging effects of lithium ion batteries by impedance spectroscopy", Electrochimica Acta, vol. 51, no. 8–9, pp. 1664–1672, 2006, doi: 10.1016/j.electacta.2005.02.148, issn: 0013-4686 (cit. on pp. 3, 159).

[25] S. Srivastav, M. J. Lacey, and D. Brandell, "State-of-charge indication in Li-ion batteries by simulated impedance spectroscopy", Journal of Applied Electrochemistry, vol. 47, no. 2, pp. 229–236, 2017.

[26] D. Stroe, M. Swierczynski, A. Stroe, and S. K. Kær, "Generalized characterization methodology for performance modelling of Lithium-Ion batteries", Batteries, vol. 2, no. 4, p. 37, 2016, doi: 10.3390/batteries2040037.

[27] T. Kazuhiko, I. Masahiro, T. Kazuo, Y. Junichi, and O. Shigeto, "Quick testing of batteries in lithium-ion battery packs with impedance-measuring technology", Journal of Power Sources, vol. 128, no. 1, pp. 67–75, 2004.

[28] A. Cuadras and O. Kanoun, "SoC Li-ion battery monitoring with impedance spectroscopy", In: 6th International Multi-Conference on Systems, Signals and Devices (SSD'09). 2009, pp. 1–5, doi: 10.1109/SSD.2009.4956761.

[29] P. J. Osswald, S. V. Erhard, A. Noel, P. Keil, F. M. Kindermann, H. Hoster, and A. Jossen, "Current density distribution in cylindrical Li-Ion cells during impedance measurements", Journal of Power Sources, vol. 314, pp. 93–101, 2016.

[30] T. Günther, P. Büschel, and O. Kanoun, "Eingebettetes Impedanzmesssystem für das Batteriemanagement in Elektrofahrzeugen", Technisches Messen, vol. 81, no. 11, pp. 560–565, 2014, doi: 10.1515/teme-2014-1052.

[31] R. R. Richardson, P. T. Ireland, and D. A. Howey, "Battery internal temperature estimation by combined impedance and surface temperature measurement", Journal of Power Sources, vol. 265, pp. 254–261, Nov. 2014, doi: 10.1016/j.jpowsour.2014.04.129.

[32] J. Wu, C. Zhang, and Z. Chen, "An online method for lithium-ion battery remaining useful life estimation using importance sampling and neural networks", Applied Energy, vol. 173, pp. 134–140, 2016.

[33] U. G. Kyle, I. Bosaeus, A. D. De Lorenzo, P. Deurenberg, M. Elia, J. M. Gómez, B. L. Heitmann, L. Kent-Smith, J.-C. Melchior, M. Pirlich, H. Scharfetter, A. M. W. J. Schols, and C. Pichard, "Bioelectrical impedance analysis - Part I: Review of principles and methods", Clinical Nutrition, vol. 23, no. 5, pp. 1226–1243, October 2004.

[34] U. M. Moissl, P. Wabel, P. W. Chamney, I. Bosaeus, N. W. Levin, A. Bosy-Westphal, O. Korth, M. J. Müller, L. Ellegård, V. Malmros, C. Kaitwatcharachai, M. K. Kuhlmann, F. Zhu, and N. Fuller, "Body fluid volume determination via body composition spectroscopy in health and disease", Physiological Measurement, vol. 27, no. 9, pp. 921–933, Article number 012, 1 September 2006.

[35] M. Min and T. Parve, "Improvement of lock-in electrical bio-impedance analyzer for implantable medical devices: Contactless sensing of the conductivity of aqueous droplets in segmented flow", IEEE Transactions on Instrumentation and Measurement, vol. 56, no. 3, pp. 968–974, 2007.

[36] B. Sanchez, G. Vandersteen, I. Martin, D. Castillo, A. Torrego, P. J. Riu, J. Schoukens, and R. Bragos, "In vivo electrical bioimpedance characterization of human lung tissue during the

bronchoscopy procedure. A feasibility study", Medical Engineering and Physics, vol. 35, no. 7, pp. 949–957, 2013.

[37] H. Morgan, T. Sun, D. Holmes, S. Gawad, and N. G. Green, "Single cell dielectric spectroscopy", Journal of Physics D, Applied Physics, vol. 40, no. 1, S10, pp. 61–70, 2006.

[38] U. Pliquett, "Bioimpedance: A review for food processing", Food Engineering Reviews, vol. 2, no. 2, pp. 74–94, June 2010, doi: 10.1007/s12393-010-9019-z.

[39] M. Guermazi, U. Tröltzsch, O. Kanoun, and N. Derbel, "Influence of meat aging on modified Fricke model's parameter", In: O. Kanoun, Lecture Notes on Impedance Spectroscopy, vol. 3. Taylor and Francis, 2012.

[40] A. Chowdhury, T. Kanti Bera, D. Ghoshal, and B. Chakraborty, "Electrical impedance variations in banana ripening: An analytical study with electrical impedance spectroscopy", Journal of Food Process Engineering, vol. 40, no. 2, Article number e12387, 1 April 2017, doi: 10.1111/jfpe.12387.

[41] A. Abdelkafi, P. Büschel, A. Fendri, and O. Kanoun, "Impedance investigation of milk dilution", In: AMA Sensor and Test, 19-21 May 2015, Nuremberg, Germany, pp. 156–159, doi: 10.5162/sensor2015/A7.2, 2015.

[42] M. Guermazi, U. Tröltzsch, O. Kanoun, and N. Derbel, "Assessment of beef meat aging using impedance spectroscopy", In: 8th International Multi-Conference on Systems, Signals and Devices, Sousse, Tunisia. 22–25 March 2011, pp. 1–6, doi: 10.1109/SSD.2011.5767450.

[43] M. Heidary Dastjerdi, O. Kanoun, and J. Himmel, "Verfahren zum Abgleich von Gradiometern für medizinische Anwendungen", Technisches Messen, vol. 83, no. 5, pp. 247–256, 2016.

[44] D. Bouchaala, O. Kanoun, and N. Derbel, "High accurate and wideband current excitation for bioimpedance health monitoring systems", Measurement, vol. 79, pp. 339–348, February 2016.

[45] B. H. Brown, J. A. Tidy, K. Boston, A. D. Blacket, R. H. Smallwood, and F. Sharp, "Relation between tissue structure and imposed electrical current flow in cervical neoplasia", Lancet, 2000, pp. 892–895, doi: 10.1016/S0140-6736(99)09095-9.

[46] D. Bouchaala, M. Guermazi, N. Derbel, and O. Kanoun, "Portable device design for in-vitro muscle tissue monitoring", tm – Technisches Messen, vol. 82, no. 10, pp. 485–494, 2015.

[47] M. Min, U. Pliquett, T. Nacke, A. Barthel, P. Annus, and R. Land, "Broadband excitation for short-time impedance spectroscopy", Physiological Measurement, vol. 29, no. 6, pp. 185–192, 2008.

[48] Y. Yang, F. Zhang, K. Tao, L. Wang, H. Wen, and Z. Teng, "Multi-frequency simultaneous measurement of bioimpedance spectroscopy based on a low crest factor multisine excitation", Physiological Measurement, vol. 36, no. 3, pp. 489–501, 1 March 2015.

[49] D. Bouchaala, E. Mekki, T. Guenther, P. Bueschel, O. Kanoun, and N. Derbel, "Study of excitation signals parameters for portable biomedical devices", In: IEEE International Instrumentation and Measurement Technology Conference, 11–14 May 2015, Pisa, Italy, 2015, pp. 784–788.

[50] B. Sanchez, E. Louarroudi, E. Jorge, J. Cinca, R. Bragos, and R. Pintelon, "A new measuring and identification approach for time-varying bioimpedance using multisine electrical impedance spectroscopy", Physiological Measurement, vol. 34, no. 3, pp. 339–357, March 2013.

[51] H.-R. Tränkler, O. Kanoun, and D. Pawelczak, "Technology, signal processing and applications: Charm of sensors", In: IEEE International Conference on Systems, Signals and Devices (SSD'07), 19–22 März 2007, Hammamet, Tunisia.

[52] O. Kanoun and H.-R. Tränkler, "Sensor technology advances and future trends", IEEE Transactions on Instrumentation and Measurement, ISSN 0018-9456, vol. 53, no. 6, 2004.

[53] H.-R. Tränkler, O. Kanoun, M. Min, and M. Rist, "Smart sensor systems using impedance Spectroscopy", Proceedings of the Estonian Academy of Sciences, vol. 13, no. 4, pp. 455–478, 2007, ISSN 1406-0175.
[54] O. Kanoun, "Untersuchungen zu Sensoren mit variierter Anregung", Technisches Messen, vol. 73, no. 7/8, pp. 382–392, 2006, 69(4/2002), doi: 10.1524/teme.2002.69.4.187.
[55] H.-R. Tränkler and O. Kanoun, "Improvement of sensor information using multi-sensor and model-based sensor systems", In: IEEE Instrumentation and Measurement Technology Conference, pp. 2259–2263, 17–19 Mai 2005, Ottawa, Ontario, Canada.
[56] T. Morgenstern, R. Brodskiy, J. Weidenmüller, O. Kanoun, and J. Himmel, "3D Eddy current modelling for non destructive crack detection", Lecture Notes on Impedance Spectroscopy, vol. 5: Taylor and Francis, 2015, pp. 65–71.
[57] M. Heidary Dastjerdi, O. Kanoun, D. Rüter, K. Thelen, J. Weidenmüller, and J. Himmel, "Magnetic field topologies for fast impedance mapping of biological tissue", In: International Workshop on Impedance Spectroscopy (IWIS), 26–28 September 2012, Chemnitz, p. 52.
[58] D. Grieshaber, R. MacKenzie, J. Vörös, and E. Reimhult, "Electrochemical biosensors - Sensor principles and architectures", Sensors, vol. 8, no. 3, pp. 1400–1458, March 2008.
[59] E. B. Bahadır and M. K. Sezgintürk, "A review on impedimetric biosensors", Artificial Cells, Nanomedicine, and Biotechnology, vol. 44, no. 1, pp. 248–262, 2016, doi: 10.3109/21691401.2014.942456.
[60] J. Munoz, R. Montes, and M. Baeza, "Trends in electrochemical impedance spectroscopy involving nanocomposite transducers: Characterization, architecture surface and bio-sensing", Trends in Analytical Chemistry, vol. 97, pp. 201–215, 2017.
[61] M. Hnaien, F. Lagarde, J. Bausells, A. Errachid, and N. Jaffrezic-Renault, "A new bacterial biosensor for trichloroethylene detection based on a three-dimensional carbon nanotubes bioarchitecture", Analytical and Bioanalytical Chemistry, vol. 400, no. 4, pp. 1083–1092, May 2011.
[62] R. Pruna, F. Palacio, A. Baraket, N. Zine, A. Streklas, J. Bausells, A. Errachid, and M. López, "A low-cost and miniaturized potentiostat for sensing of biomolecular species such as TNF-α by electrochemical impedance spectroscopy", Biosensors and Bioelectronics, vol. 100, pp. 533–540, 15 February 2018.
[63] I. O. K'Owino and O. A. Sadik, "Impedance Spectroscopy: A powerful tool for rapid biomolecular screening and cell culture monitoring", Electroanalysis, vol. 17, no. 23, pp. 2101–2113, 2005, doi: 10.1002/elan.200503371.
[64] S. Z. Mousavisani, J.-B. Raoof, A. P. F. Turner, R. Ojani, and W. C. Mak, "Label-free DNA sensor based on diazonium immobilisation for detection of DNA damage in breast cancer 1 gene", Sensors and Actuators B, Chemical, vol. 264, pp. 59–66, 1 July 2018.
[65] B. Wang, U. Akiba, and J. Anzai, "Recent progress in nanomaterial-based electrochemical biosensors for cancer biomarkers: A review", Molecules, vol. 22, no. 7, 2017.

Part I: Electrochemical and physical phenomena

Abdulkadir Sanli, Abderrahmane Benchirouf, Christian Müller, and Olfa Kanoun

Study of the humidity effect on the electrical impedance of MWCNT epoxy nanocomposites

Abstract: Owing to the excellent physical properties of carbon nanotubes (CNTs), incorporation of CNTs into polymer matrices would make the nanocomposites a potential candidate for smart sensing applications. However, environmental factors such as humidity might influence the nanocomposite properties and performance. In this work, the effect of humidity on the electrical properties of multiwalled CNT (MWCNT) epoxy nanocomposites is investigated by electrochemical impedance spectroscopy in the frequency range from 40 Hz to 110 MHz. The MWCNT epoxy nanocomposites with different MWCNT concentrations up to 1 % wt were fabricated by the direct mixing method and the composites were deposited on the glass epoxy substrate by the stencil printing technique. From the impedance results, it was observed that the overall impedance of the nanocomposites tends to increase with the increase of the humidity due to an increase in tunneling barriers induced by swelling of the polymer layer between the CNT junctions within the nanocomposite matrix. The humidity effect tends to decrease with the increase of MWCNT concentration due to a decrease of the tunneling effect at high MWCNT concentrations. Furthermore, the proposed nanocomposites at a 0.5 % wt concentration show a humidity sensitivity of about 0.006 % relative humidity (RH) and 0.004 % RH and linear responses with linear correlations (R^2) of 0.98 and 0.93 for 100 Hz and 1 kHz, respectively. Moreover, the sensors were covered with latex polymer to minimize the humidity effects. It was found that with encapsulation of the nanocomposites, the relative impedance change of the nanocomposite is decreased from 3.76 % to 1.15 % at a 0.5 % wt MWCNT concentration.

Keywords: carbon nanotubes, epoxy resin, nanocomposites, humidity sensing, encapsulation, electrochemical impedance spectroscopy

1 Introduction

Since their discovery of by Iijima in 1991 [1], carbon nanotubes (CNTs) have been used extensively in most fields of science and engineering owing to their excellent mechan-

Acknowledgement: This work was partially supported by the Ministry of National Education of Turkey, which provided Abdulkadir Sanli with a scholarship. The authors gratefully acknowledge Mr. Ammar Al-Hamry for designing the salt solutions for humidity investigations.

Abdulkadir Sanli, Abderrahmane Benchirouf, Christian Müller, Olfa Kanoun, Chair for Measurement and Sensor Technology, Chemnitz University of Technology, Reichenhainer Strasse 70, 09126 Chemnitz, Germany, e-mail: abdulkadir.sanli@s2012.tu-chemnitz.de

https://doi.org/10.1515/9783110558920-002

ical, thermal and electrical properties [2–5]. In comparison to other fillers, such as nanoparticles or carbon black, CNTs are considered as an effective filler owing to their high aspect ratios (≈ 100–1000), which result in the formation of conductive paths at very low CNT concentrations [6, 7]. Thus, integration of CNTs into a polymer matrix would make the composites a potential candidate for smart sensing applications, including chemical, pressure, gas, flow and mechanical sensors [8–11]. The environmental factors, e. g., humidity, can influence the overall electrical conductivity as well as sensor performance of a polymer nanocomposite for the sensor applications such as strain and pressure sensing, in which the compensation of environmental effects is required. Therefore, it is important to quantify the influence of environmental factors on the electrical conductivity of CNT-based nanocomposites. To reveal the conduction mechanism of the CNT polymer nanocomposites under the humidity effects, several researchers have proposed various explanations [12–17]. For instance, Yoo et al. [13] suggested that resistivity changes are caused by polymer swelling below the percolation threshold, whereas above the percolation threshold the main reason in resistivity change is the charge transfer from water molecules to CNTs. On the other hand, Yu et al. [14] suggested that the change in the conductivity is due to the formation of a weak bond between H and C atoms from water and CNTs induced by water uptake. Asrafi et al. [16] have found that a change of humidity from 5 % relative humidity (RH) to 95 % RH can cause an 80 % reduction in conductivity of CNT epoxy nanocomposites due to moisture ingress. However, to the best of our knowledge, there is no study addressing the effects of the CNT concentration and polymer coating on the electrical properties of MWCNT epoxy nanocomposites under humidity effects.

Therefore, this work focuses on the quantitative investigation of the effect of humidity on the electrical conductivity of a free standing and encapsulated MWCNT epoxy nanocomposite by electrochemical impedance spectroscopy for different MWCNT concentrations.

2 Experimental procedures

2.1 Measurement setup for humidity investigations

Prior to placing the samples inside the salt container, a precise humidity sensor is used to identify the values of humidity and temperature within the salt container. After the sample has been placed inside the container, impedance measurements were taken after ca. 45 min for the stabilization of the humidity. After that, the impedance data were recorded by utilizing the Agilent 4294A precision impedance analyzer in the frequency range of 40 Hz to 110 MHz. The measurement setup for humidity analysis of MWCNT epoxy nanocomposites is represented in Fig. 1.

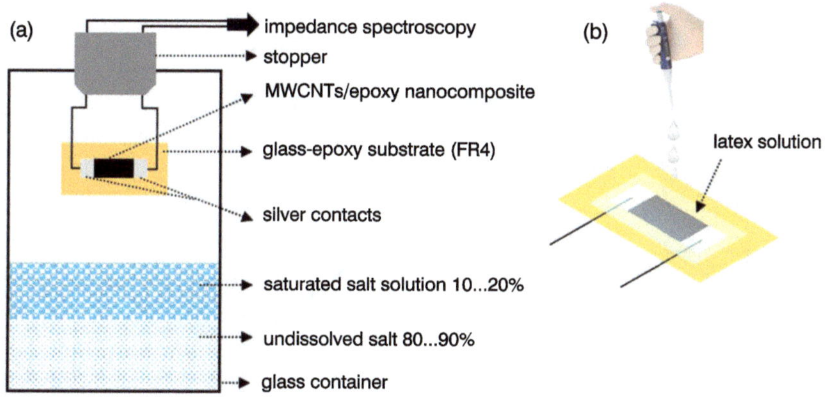

Fig. 1: Schematic illustration of (a) humidity measurement setup for humidity investigations of MWCNT epoxy nanocomposites placed inside a glass container and (b) encapsulation of samples with latex solution.

3 Results and discussion

3.1 Effect of humidity on the impedance response

It can be seen from both the Nyquist plot and the Bode plot (see Fig. 2) that the impedance of the MWCNT epoxy nanocomposites tend to increase with the increase of RH. The Nyquist plot shows quasi-semi-circle behavior for all concentrations and humidity responses. In order to estimate the complex transport mechanism in MWCNT epoxy nanocomposites under the humidity effect, a 3D R-C model can be proposed to reveal the effect of the MWCNT concentration and the interaction within the nanocom-

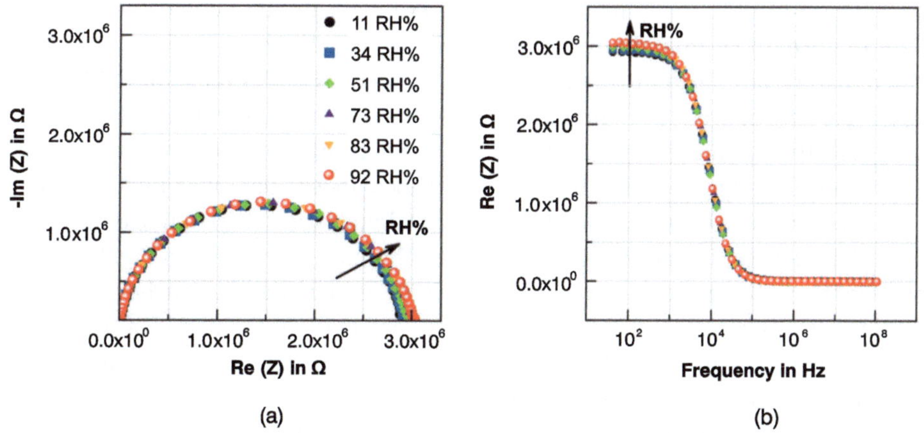

Fig. 2: (a) Nyquist and (b) Bode plot of non-capsulated MWCNT epoxy nanocomposites under RH.

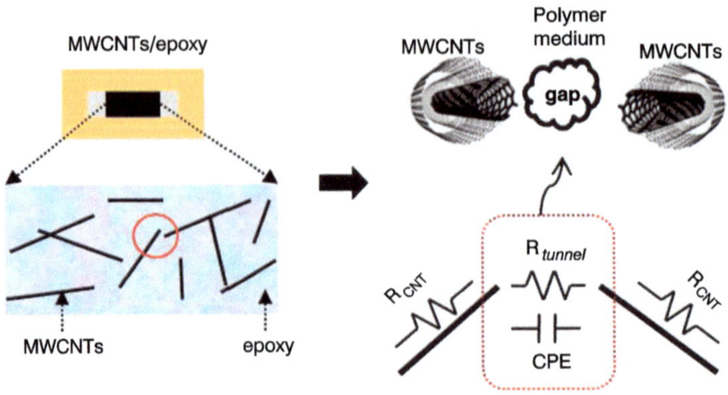

Fig. 3: Illustration of the working mechanism and corresponding equivalent circuit model of an MWCNT epoxy nanocomposite.

posite matrix, shown in Fig. 3. Here, each nanotube (or nanotube agglomerate) has a resistance R_{CNTs} and the polymer medium can be modeled as R_{tunnel} and constant phase element (CPE), respectively.

3.1.1 Humidity sensing mechanism

From the accumulated studies [18–23], the effect of humidity on the CNT polymer nanocomposites can be attributed to the following working mechanisms: (i) the increase of intrinsic resistance of CNTs due to adsorption of water molecules [20], (ii) the charge transfer between the water molecules and CNT modules and the Fermi level that causes the resistance change across CNT–CNT junctions [21, 22] and (iii) the change in inter-tube distance induced by polymer swelling under the humidity [23]. For the case of CNT epoxy nanocomposites, the main reason behind the increase of impedance under humidity is the polymer swelling. Many researchers have suggested that the swelling of the polymer layer between CNT–CNT junctions within the nanocomposite matrix due to humidity causes the increase of the tunneling barriers [18, 23]. As the polymer expands, the CNTs in the network within the polymer matrix tend to move apart from each other as if the nanocomposite deformed under the tension, which causes the alteration of the complex impedance of the nanocomposites. Thus, the electrical impedance increases under humidity as shown in Fig. 4. Therefore, the effect of intrinsic resistance change of CNTs under humidity can be neglected.

Here, it was also found that the humidity sensitivity of the nanocomposite decreases with increasing the MWCNT concentration due to a decrease in the effect of the tunneling resistance. When the CNT concentration is high, the MWCNTs locate very close within the nanocomposite matrix. Therefore, the inter-tube distances at high concentrations are less effected by the polymer swelling than those near the perco-

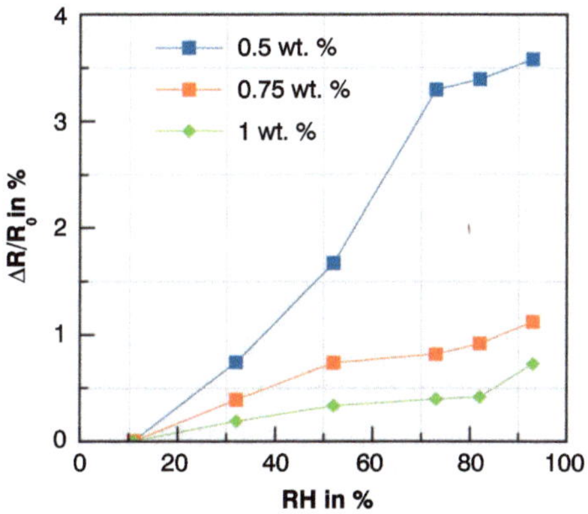

Fig. 4: Relative resistance chance of the MWCNT epoxy nanocomposites under RH for different MWCNT concentrations.

lation threshold [19]. While for the samples of a 0.5 % wt MWCNT concentration the relative resistance change is 3.76 %, this value decreases to 0.72 % for 1 % wt MWCNTs. Moreover, as illustrated in Fig. 5, it was seen that the operation frequency has a great effect on the impedance in MWCNT epoxy nanocomposites under RH. It has been observed that the proposed nanocomposites of a 0.5 wt % MWCNT concentration show humidity sensitivities of about 0.006 % RH and 0.004 % RH with linear responses (R^2) of 0.98 and 0.93 for 100 Hz and 1 kHz, respectively. The proposed MWCNT

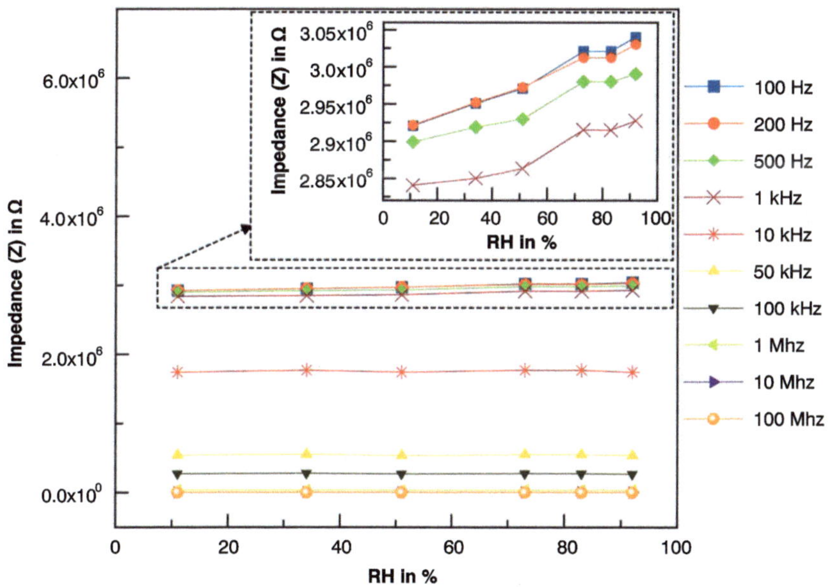

Fig. 5: Impedance response of MWCNT epoxy nanocomposite at a 0.5 % wt MWCNT concentration under RH for different frequencies up to 100 MHz. Inset image shows the corresponding impedance response of the MWCNT epoxy nanocomposites up to 1 kHz.

epoxy nanocomposite can be also used as humidity sensor as it was found to be more sensitive than many other nanocomposites [13–24].

3.2 Effect of the encapsulation on the impedance response

In order to minimize the effect of the humidity on the electrical properties of MWCNT epoxy nanocomposites, samples were encapsulated with latex polymer (see Fig. 6b). After that, the samples underwent the same humidity measurements again. From Fig. 6, it can be seen that the resistance value of the composites slightly decreases after the encapsulation due to the effect of the latex polymer, which acts as an additional parallel resistance to the composite. By extracting the resistance values of the encapsulated nanocomposites, it can be seen that the latex encapsulation significantly minimizes the effect of the humidity. While the relative resistance change of the nanocomposites for 0.5 % wt decreases from 3.76 % to 1.15 %, this value decreases from 0.72 % to 0.52 % for a 1 % wt MWCNT concentration, as shown in Fig. 7.

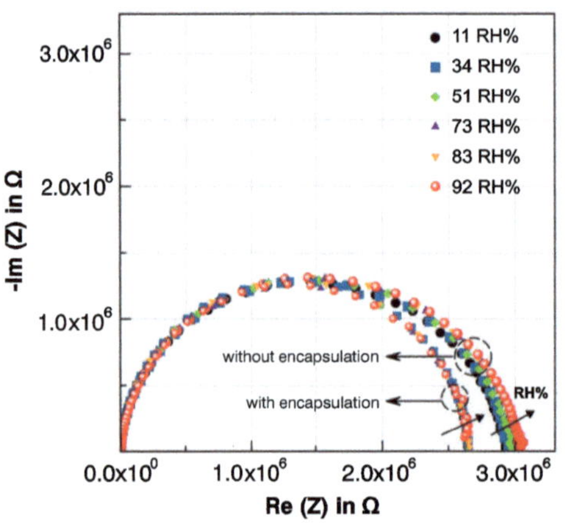

Fig. 6: Nyquist plot of encapsulated and non-encapsulated MWCNT epoxy nanocomposites at a 0.5 % wt MWCNT concentration under up to 92 % RH.

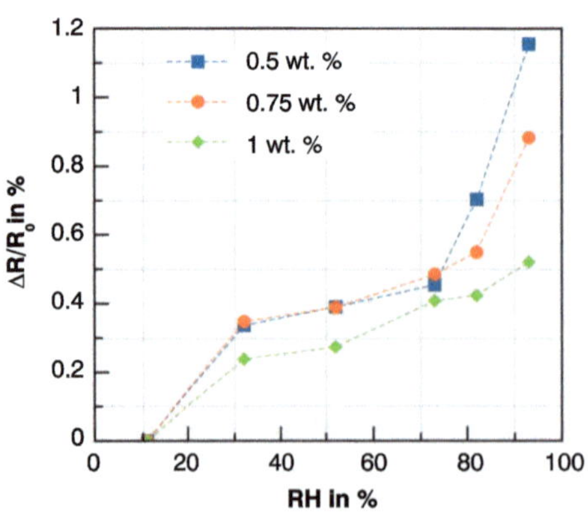

Fig. 7: Relative resistance chance of the latex-encapsulated MWCNT epoxy nanocomposites under RH for different MWCNT concentrations.

4 Conclusions and outlook

In this study, we investigated the effect of humidity on the impedance properties of MWCNT epoxy nanocomposites for different MWCNT concentrations. From the impedance results, we found that the impedance of the nanocomposites increases with the increase of the humidity due to an increase in tunneling barriers induced by swelling of the polymer layer. Additionally, the humidity effect was found to decrease with the increase of MWCNT concentration due to the decrease of the tunneling effect at high MWCNT concentrations. Furthermore, the proposed nanocomposites at a 0.5 wt % MWCNT concentration show a humidity sensitivity of about 0.006 % RH and a linear response (R^2) of 0.98. Moreover, by encapsulation of the nanocomposites the relative impedance change of the nanocomposite decreases from 3.76 % to 1.15 % at a 0.5 % wt MWCNT concentration. For the future work, it would be interesting to explore the effect of the different encapsulation methods on the electrical and piezoresistive properties of the MWCNT epoxy nanocomposites.

Bibliography

[1] S. Bianco, C. F. Pirri, M. Quaglio, P. Ferrario, and R. Castagna, "Nanocomposites based on elastomeric matrix filled with carbon nanotubes for biological applications", INTECH Open Access Publisher.

[2] Z. Yao, C. L. Kane, and C. Dekker, "High-field electrical transport in single-wall carbon nanotubes", Physical Review Letters, vol. 84, no. 13, p. 2941.

[3] P. R. Bandaru, "Electrical properties and applications of carbon nanotube structures", Journal of Nanoscience and Nanotechnology, vol. 7, no. 4–5, pp. 1239–1267.

[4] M. J. Treacy, T. W. Ebbesen, and J. M. Gibson, "Exceptionally high Young's modulus observed for individual carbon nanotubes".

[5] M. F. Yu, B. S. Files, S. Arepalli, and R. S. Ruoff, "Tensile loading of ropes of single wall carbon nanotubes and their mechanical properties", Physical Review Letters, vol. 84, no. 24, p. 5552.

[6] J. K. W. Sandler, S. Pegel, M. Cadek, F. Gojny, M. Van Es, J. Lohmar, and M. S. P. Shaffer, "A comparative study of melt spun polyamide-12 fibres reinforced with carbon nanotubes and nanofibres", Polymer, vol. 45, no. 6, pp. 2001–2015.

[7] D. S. McLachlan, C. Chiteme, C. Park, K. E. Wise, S. E. Lowther, P. T. Lillehei, and J. S. Harrison, "AC and DC percolative conductivity of single wall carbon nanotube polymer composites", Journal of Polymer Science Part B, Polymer Physics, vol. 43, no. 22, pp. 3273–3287.

[8] O. Kanoun, C. Müller, A. Benchirouf, A. Sanli, T. N. Dinh, A. Al-Hamry, and A. Bouhamed, "Flexible carbon nanotube films for high performance strain sensors", Sensors, vol. 14, no. 6, pp. 10042–10071.

[9] A. Benchirouf, S. Palaniyappan, R. Ramalingame, P. Raghunandan, T. Jagemann, C. Müller, and O. Kanoun, "Electrical properties of multi-walled carbon nanotubes/PEDOT: PSS nanocomposites thin films under temperature and humidity effects", Sensors and Actuators B, Chemical, vol. 224, pp. 344–350.

[10] O. Kanoun, C. Müller, A. Benchirouf, A. Sanli, A. Bouhamed, A. Al-Hamry, and L. Bu, "Potential of flexible carbon nanotube films for high performance strain and pressure sensors", Nanotechnology for Optics and Sensors, One Central (OCP).
[11] A.Sanli, C. Müller, O. Kanoun, C. Elibol, and M. F. X. Wagner, "Piezoresistive characterization of multi-walled carbon nanotube-epoxy based flexible strain sensitive films by impedance spectroscopy", Composites Science and Technology, vol. 122, pp. 18–26.
[12] T. Fei, K. Jiang, F. Jiang, R. Mu, and t. Zhang, "Humidity switching properties of sensors based on multiwalled carbon nanotubes/polyvinyl alcohol composite films", Journal of Applied Polymer Science, vol. 131, no. 1.
[13] K. P. Yoo, L. T. Lim, N. K. Min, M. J. Lee, C. J. Lee, and C. W. Park, "Novel resistive-type humidity sensor based on multiwall carbon nanotube/polyimide composite films", Sensors and Actuators B, Chemical, vol. 145, no. 1, pp. 120–125.
[14] H. Yu, T. Cao, L. Zhou, E. Gu, D. Yu, and D. Jiang, "Layer-by-layer assembly and humidity sensitive behavior of poly (ethyleneimine)/multiwall carbon nanotube composite films", Sensors and Actuators B, Chemical, vol. 119, no. 2, pp. 512–515.
[15] P. G. Su and C. S. Wang, "In situ synthesized composite thin films of MWCNTs/PMMA doped with KOH as a resistive humidity sensor", Sensors and Actuators B, Chemical, vol. 124, no. 2, pp. 303–308.
[16] Q. Y. Tang, Y. C. Chan, and K. Zhang, "Fast response resistive humidity sensitivity of polyimide/multiwall carbon nanotube composite films", Sensors and Actuators B, Chemical, vol. 152, no. 1, pp. 99–106.
[17] B. Ashrafi, D. Marchand, Y. Martinez-Rubi, M. B. Jakubinek, B. Simard, and A. Johnston, "Effect of Humidity on Electrical conductivity of Carbon Nanotube-Modified Epoxy".
[18] L. Liu, X. Ye, K. Wu, Z. Zhou, D. Lee, and T. Cui, "Humidity sensitivity of carbon nanotube and poly (dimethyldiallylammonium chloride) composite films", IEEE Sensors Journal, vol. 9, no. 10, pp. 1308–1314.
[19] H. Lei, W. G. Pitt, L. K. McGrath, and C. K. Ho, "Modeling carbon black/polymer composite sensors", Sensors and Actuators B, Chemical, vol. 125, no. 2, pp. 396–407.
[20] A. Zahab, L. Spina, P. Poncharal, and C. Marliere, "Water-vapor effect on the electrical conductivity of a single-walled carbon nanotube mat", Physical Review B, vol. 62, no. 15, p. 10000.
[21] J. Li, Y. Lu, Q. Ye, M. Cinke, J. Han, and M. Meyyappan, "Carbon nanotube sensors for gas and organic vapor detection", Nano Letters, vol. 3, no. 7, pp. 929–933.
[22] J. Zhao, A. Buldum, J. Han, and J. P. Lu, "Gas molecule adsorption in carbon nanotubes and nanotube bundles", Nanotechnology, vol. 13, no. 2, p. 195.
[23] D. Sung, S. Hong, Y. Kim, N. Park, S. Kim, S. L. Maeng, and K. C Kim, "Ab initio study of the effect of water adsorption on the carbon nanotube field-effect transistor", Applied Physics Letters, vol. 89, no. 24, p. 243110.
[24] L. Liu, X. Ye, K. Wu, R. Han, Z. Zhou, and T. Cui, "Humidity sensitivity of multi-walled carbon nanotube networks deposited by dielectrophoresis", Sensors, vol. 9, no. 3, pp. 1714–1721, 2009.

Lukasz Macioszek and Ryszard Rybski

Temperature influence on impedance of premium summer diesel fuel measured with the use of impedance spectroscopy

Abstract: In this chapter, an attempt is made to experimentally verify temperature influence on the evaluation of diesel fuel impedance with the use of electrochemical impedance spectroscopy. Previous test findings showed that this method can be used to detect contamination, such as water, in constant temperature. Tests presented in this chapter show that temperature itself can change electrical properties of diesel fuel similarly to water. Moreover, tested premium diesel, as a function of temperature, behaved oppositely from what hydrocarbons normally do. This makes temperature and additives present in fuel very important factors, which must be taken into consideration in the evaluation of any diesel properties using impedance spectroscopy.

Keywords: diesel, fuel, impedance spectroscopy, temperature

1 Introduction

Diesel is a very common fuel used to drive engines of vehicles, locomotives, ships and even electricity generation plants. It is a result of crude oil distillation and consists of several types of hydrocarbons (about 75 % saturated, 25 % aromatic), as well as various kinds of additives and enhancers added during or after production [1]. A few percent of fatty acid methyl ester (FAME) is also added in many countries to each fuel commercially available, for example, in Poland there is 7 % (V/V) of biocomponents.

Diesel analysis is very difficult in terms of interpreting obtained results. The encountered problems can be divided into two basic kinds. The first is that only very few research institutes are studying electrical properties of fuel, including its impedance. They have to be studied at refinery laboratories but the results are not publicly available. In such a case, information can only be obtained via private communication with people working in those facilities. The publicly available publications are mainly focused on fossil diesel rather than modern commercially sold fuel. The second kind of problems for a researcher is connected to the fact that the precise chemical composition of diesel is never the same. Responsible for this is the distillation, which, like any process, is never performed with 100 % efficiency. Additionally, the crude oil itself is not always the same. There are also a number of additives that make fuel safe to

Lukasz Macioszek, Ryszard Rybski, Institute of Metrology, Electronics and Computer Science, Faculty of Computer, Electrical and Control Engineering, University of Zielona Gora, Zielona, Poland, e-mail: l.macioszek@imei.uz.zgora.pl

https://doi.org/10.1515/9783110558920-003

transport, improve its lubricity, reduce the risk of freezing in winter, etc. [2]. Used additives are always producers' secrets and may change at any time, even though there is an obvious need for a maintenance of the quality of diesel fuel that is being sold. Within the European Union the EN 590 standard is valid. It contains ranges of physical properties' permissible values and lists the corresponding test methods to verify them [3].

Premium-type diesel fuels are declared by producers as containing engine cleaning substances [4], improving lubricity [5], having a higher cetane number [6, 7] and being sulphur-free [7]. Some are advertised as being capable of protecting engines against corrosion and increasing the driving range in comparison to regular fuel [5]. Winter premium diesel should enable proper functioning of engines in lower ambient temperatures than regular diesel. Due to this fact it has to contain additional substances. Aforementioned characteristics make the analysis of premium-type fuel even more complicated. Nevertheless, premium diesel is believed to have more consistent content and this is the reason why it was chosen to be examined.

The improved efficiency and performance of diesel engines imply an increase in their sensitivity to all kinds of contaminants in the fuel. Very high pressures, exceeding 300 MPa, are featured by modern direct injection systems (called common rail). Such values would not be possible to achieve without rigorous fuel parameters that ensure the absence of substances aggressive towards metal alloys, of which engines are composed. Particularly newer engines' fuel rails are especially sensitive to water content in the fuel; therefore introducing additional contamination measurement methods seems to be necessary. Only 200 mg/kg of water is allowed by the aforementioned standard EN 590. The coulometric Karl Fischer titration method is used for this determination, as it allows obtaining precise results in the range of 0.003–0.100 % (m/m). The solubility limit of water in diesel is reported to be very low (100 ppm at 40°C [8]) but various kinds of additives can make it significantly higher [2]. Although amounts found in fuel can be considered as relatively low, water should still be treated as a dangerous contamination. Coulometric titration is a very precise method, though it cannot be used *in situ* as it needs to be performed in the laboratory and requires preparation of test samples. One of the various methods proposed for the evaluation of diesel fuel properties is impedance spectroscopy [9, 10] but it has not yet been used for water content assessment in any device. Such device, if designed as portable, could be used for example at the gas station to perform a quick preliminary measurement and indicate whether the examined fuel may be contaminated and should be sent to the laboratory for further, more precise studies.

In order to propose a complete method of measuring water content as well as any other contaminant in diesel fuel with the use of impedance spectroscopy, evaluation of the influence of the temperature on diesel impedance is required. Pure diesel fuel can be treated as dielectric [9] but each additive makes its dielectric properties more complex. Hydrocarbons, of which diesel is composed, tend to lower their resistivity with the raise of temperature. Since around 2006 the production of ultra-low sulphur diesel (ULSD) has been growing and this kind of fuel dominates the market nowadays. Processes used to remove sulphur also significantly decrease fuel conductivity [2]. On the other hand, transportation standards require the conductivity of diesel to be above 50 pS/m [11] to allow static charge to dissipate. This is the reason for adding substances that will not noticeably change all fuel properties except conductivity. What kind of additives are added is a secret of the fuel producer. There are many substances available but their composition is also classified in general. It is known that additives are added in very small proportions, of a few tens to hundreds parts per million, and they can raise diesel conductivity tens of times [2]. This implies that additives may alter or overshadow the behavior of hydrocarbons impedance as a function of temperature, thus measurements of diesel impedance at different temperatures are very important to find their correlation. Once this correlation is found, it should be included in the diesel electric equivalent circuit in order to model fuel behavior at different temperatures. Such a model is necessary to assess fuel contamination with the use of a portable device based on impedance spectroscopy in the future.

2 Material and methods

The main objective of this study is to experimentally check the temperature influence on the impedance of premium summer diesel fuel measured with the use of impedance spectroscopy. Fuel samples were obtained from a gas station and were examined at different temperatures in a temperature controlling container. Impedance values were the same during measurement series with increasing and decreasing temperature. Once impedance of samples was measured as a function of frequency, the results were used to fit values of diesel equivalent circuit elements.

2.1 Diesel fuel samples

Diesel samples of 500 ml capacity were examined in glass beakers (shown in Fig. 1 (a)). This rather high capacity was used in previous experiments to prevent unplanned, spontaneous changes of temperature during the experiment. Although such situations did not take place, it was used again to preserve maximum possible repeatability of experiments. Temperature was monitored all the time and was approximately constant (±0.1°C) during each experiment. Measurements were performed at 6.9°C, 9.3°C, 11.1°C, 14.5°C, 16.4°C, 18.4°C, 23°C and 26°C (±0.1°C). These values were chosen arbitrarily as preliminary ones just to cover the basic range. With the equipment the used temperature could be more precisely measured than be set; nevertheless, the measuring conditions were good enough to perform planned studies on the behavior of diesel impedance at different temperatures.

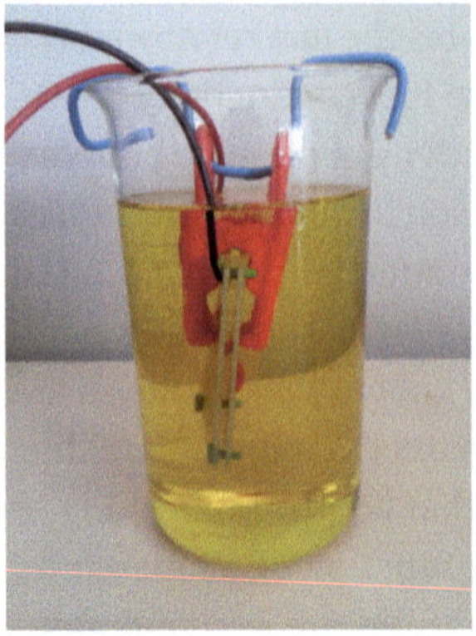

(a)

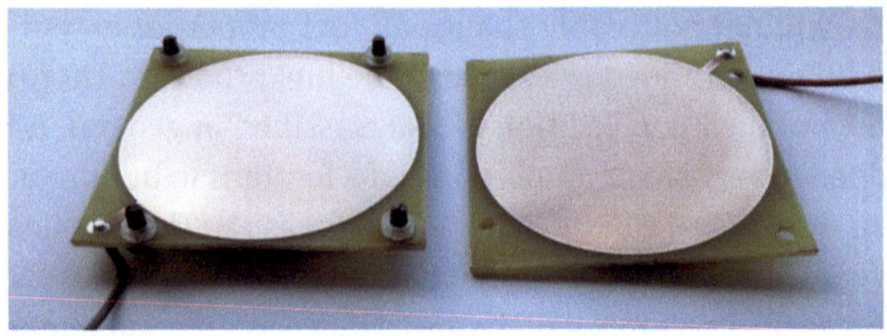

(b)

Fig. 1: (a) Sample with immersed electrodes. (b) Electrodes used in the experiment.

2.2 Electrodes

Measurements were performed using two circular parallel electrodes made of gold-coated copper with a surface area of 25 cm^2 each, that were spaced by 2 mm. During experiments they were immersed in diesel samples. Fig. 1 (b) illustrates both electrodes and four supports providing the planned distance. As can be seen, the actual electrodes were in the form of a copper layer (35 µm) on a laminate, the same that is usually used to build single-layer electronic circuits. Leaving gold-coated electrodes immersed in diesel for long periods of time (up to 14 days) did not reveal any chemical reactions that would take place on their surface. This allowed for the assumption that cleaning electrodes after each measurement is not necessary to maintain repeatability, which was confirmed by later obtained results.

2.3 Impedance spectroscopy

Diesel fuel sample impedance was measured in the frequency range of 0.1–500 Hz with a 700-mV RMS voltage. A laboratory EG&G/Princeton Applied Research electrochemical impedance spectroscopy system was used, consisting of the 263A Potentiostat-Galvanostat, 5210 Dual Phase Lock-In Amplifier and PowerSINE software. The system was calibrated according to the manufacturer's recommendations and self-calibration before each measurement was also performed. Each final measuring result was an average of four separate impedance measurements that were performed by the system. This number was chosen arbitrarily and was a compromise between measurement quality and time; each experiment took about twenty minutes. Measured impedance in the form of

$$Z^*(\omega) = Z'(\omega) - jZ''(\omega) \tag{1}$$

was used to fit equivalent circuit element values [12]. The circuit was simple and contained a resistor with parallel constant phase element. Such circuit was already used elsewhere to describe diesel properties [9, 10]. Impedance of such circuit can be described by

$$Z^*(\omega) = \frac{R}{(1 + QR(j\omega)^n)}, \tag{2}$$

where Q is the admittance $1/|Z|$ at $\omega = 1\,\text{rad/s}$ and real n satisfies $0 \le n \le 1$.

3 Results and discussion

Diesel fuel samples were examined at temperatures listed in Section 2.1. Stray effects can be omitted as several measurement repeats did not reveal values exceeding the measurement error of the system used. Fig. 2 presents the Nyquist plot with the measured impedance of eight diesel samples. It can be seen that measured values form single semi-circles. The right ends of the semi-circles are a bit raised which suggests that within this low-frequency range there may be valuable information. However, further measurements at frequencies down to 1 mHz did not reveal second semi-circles and hence they were not included in the chapter. Although the used frequency range of 0.1–500 Hz may seem to be narrow, it was observed in the evaluation of another dielectric, namely transformer oil, that using a wider frequency range did not reveal more information [13]. The raise may be associated with diffusion phenomena occurring in the dielectric when a metal electrode is used [14] and it is not considered as a primary and dominant phenomenon. The centers of the semi-circles are a bit depressed, which suggests that the calculated capacitance would vary with frequency. It

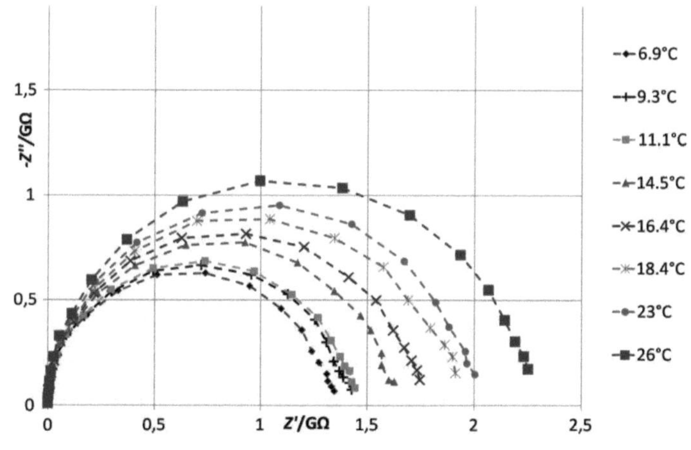

Fig. 2: Nyquist plot of examined diesel fuel samples.

confirms that choosing a simpler equivalent circuit consisting of a resistor and ideal capacitor would lead to greater fitting errors.

Note that with the raise of the temperature, the diesel resistivity increases – semi-circles have larger diameters. The real part of the measured impedance at 0.1 Hz revealed values in the range of 1.35 GΩ at 6.9°C to 2.25 GΩ at 26°C. Increasing resistivity with the raise of temperature is normally encountered in metals, and hydrocarbons should behave in the opposite way; their resistivity decreases with the increasing temperature. This may suggest that additives present in fuel responsible for increasing its conductivity play a leading role in impedance measurements. Electrode material and the fact that we used premium-type fuel were not the reasons for these unexpected results, as experiments with the use of copper electrodes and regular type diesel revealed the same resistivity behavior [15].

Since the diameters of semi-circles from the Nyquist plot were the only visible changes to their shape according to different temperatures, it was decided that overall resistance of the diesel sample can be used to propose the calibration curve that would model the temperature influence on the diesel's electrical properties. As can be seen in Fig. 3, the correlation of the samples' calculated resistance and temperature seems to be linear, with a quite good coefficient of determination ($r^2 = 0.9774$).

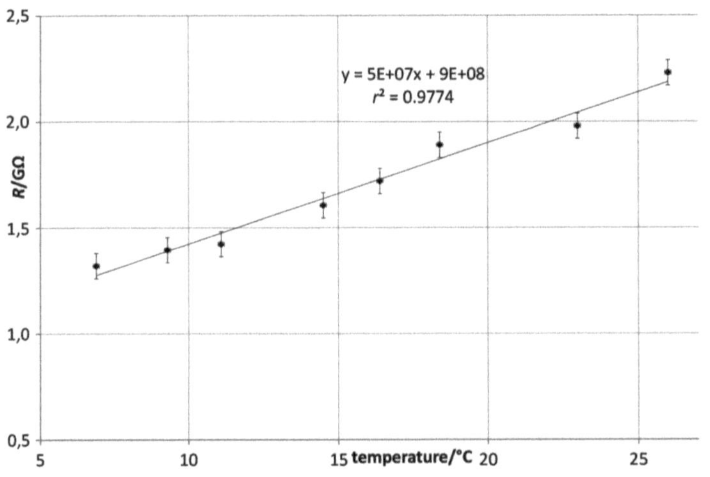

Fig. 3: Temperature dependence of fitted *R* values of studied diesel samples.

Table 1 contains fitted values of equivalent circuit elements and the corresponding χ^2 modeling errors. Relative errors of fitting individual values did not exceed 3.6 % in every examined sample. Note that the most changing value is the parallel resistivity of the equivalent circuit with 1.32 GΩ at 6.9°C and 2.229 GΩ at 26°C. Even though values of Q are also varying, it is hard to assess its trend, and therefore their differences are not

Table 1: Fitted values of equivalent circuit elements and corresponding modeling errors.

Temp. (°C)	*R* (GΩ)	*Q* (pSs^n)	*n*	Fitting (χ^2)
6.9	1.320	39.22	0.98216	0.00148
9.3	1.395	39.76	0.98600	0.00074
11.1	1.423	39.92	0.98282	0.00083
14.5	1.606	45.21	0.97990	0.00074
16.4	1.720	45.34	0.98268	0.00069
18.4	1.890	47.94	0.97489	0.00114
23	1.979	44.69	0.98356	0.00052
26	2.229	42.37	0.98278	0.00062

so significant as *R* values (up to about 20 % raise compared to 69 %) in the investigated temperature range.

A very important and unexpected result of the performed experiments is that the examined diesel fuel had a positive temperature coefficient of resistance. The experiment was repeated with the use of copper electrodes and regular (not premium) diesel. Although the fitted values of the equivalent circuit were a bit different, the examined fuel still had a positive temperature coefficient of resistance. Hydrocarbons do not have such property, thus some other substance present in the fuel caused this phenomenon. The exact chemical compositions of aforementioned additives that are used to increase the fuel's conductivity are unknown, although there is a limited number of possibilities. It is only known that these enhancers are added to fuel in proportions up to one hundred parts per million [2]. A conductivity increase of raw ULSD from about 11 pS/m to required 50 pS/m could be achieved with the addition of conductive polymers during the production. Some of these polymers can have very high conductivity, at the same level as good metal conductors. Thus, it is possible that even a very low proportion of polymers in fuel (below 50 ppm) would change conductivity of the mixture vitally. Moreover, some of the conductive polymers have a positive temperature coefficient of resistance, just like metals do. Given properties incline us to assume that it is possible that some conductive polymers could be present in the examined premium diesel fuel. Confirmation or rebuttal of this thesis will require multiple experiments with the use of different techniques, as diesel has a very heterogeneous composition. Nevertheless, an explanation of what causes diesel fuel to increase its resistivity with the raise of temperature seems to be indispensable.

The encountered behavior of diesel's impedance as a function of temperature does not mean that further studies of its impedance are pointless. Correlation between temperature and diesel's resistance in the examined range seems to be linear. This means that its influence is easy to predict and it definitely should be included in more detailed equivalent circuits that would model diesel's electric properties. Evaluation of different diesel fuel types should answer the question whether all types of diesel behave the same way, such as summer, winter, premium and regular diesel. If the temperature coefficients are, every new study of diesel's electrical properties will be valid.

4 Conclusion

In this chapter, the influence of temperature on impedance of premium summer diesel fuel measured with the use of impedance spectroscopy is presented. Experimental results are discussed and analyzed. A calibration curve as well as fitted values of the equivalent circuit are presented. They together show that temperature has a great impact on the impedance of diesel fuel.

Evaluation of the influence of temperature on diesel impedance is required in order to propose a complete method for measuring diesel contamination with the use of impedance spectroscopy. Given the limitations discussed in Section 3, we can still state that impedance spectroscopy of diesel fuel is reasonable and compares favorably with methods that require preparation of the samples and thus can only be performed in the laboratory. Analysis of the obtained results may be very difficult, as the exact composition of diesel is unknown; nevertheless, impedance spectroscopy can be used as a comparative method. As mentioned before, there does not exist nor will there ever be a reference fuel. However, changing only one property of any diesel and measuring its impedance before and after that change will always be informative. The proposed methodology may not reach the same high level of achieved accuracy as laboratory methods used nowadays; nevertheless, accuracy is not the most important factor in each case. Future studies about diesel impedance behavior in a wider temperature range with and without the addition of other substances should answer the question whether there is a possibility to indicate the harmfulness of fuel based on its measured impedance.

Advantages of using impedance spectroscopy as a measuring technique make it possible to build a device that could theoretically work *in situ*. After relatively fast measurements and proper calculations it could warn a user that contamination was found and the examined diesel fuel can be potentially dangerous to his car. If such a device would be designed to serve an inspector at a gas station, then it would indicate that the examined fuel is somehow different and it definitely should be sent to the laboratory for the whole evaluation of its properties. Plausible relatively low measurement precision of the proposed device could cause sporadic fake alarms. However, we believe that fake alarms are still far better than potentially expensive repairs of the fuel injection system in a modern car.

Bibliography

[1] Agency of Toxic Substances and Disease Registry, "Toxicological Profile for Fuel Oils", U.S. Department of Health and Human Services, Public Health Service, Atlanta, 1995.
[2] Private talk with the director of one of the Polish refinery laboratories, 2014.
[3] Automotive fuels - Diesel - Requirements and test methods, 2016.
[4] New Shell V-Power Diesel, 2016.

[5] BP Ultimate Diesel with ACTIVE technology, 2016.
[6] LOTOS Dynamic Diesel - Unique technology, 2016.
[7] VERVA ON - premium diesel, 2016.
[8] J. Van Gerpen, E. Hammond, L. Yu, and A. Monyem, "Determining the Influence of Contaminants on Biodiesel Properties", SAE Technical Paper No. 971685, Warrendale, MI: SAE, 1997.
[9] J. E. De Souza, M. D. Scherer, J. A. S. Cáceres, A. R. L. Caires, and J.-C. M'Peko, "A close dielectric spectroscopic analysis of diesel/biodiesel blends and potential dielectric approaches for biodiesel content assessment", Fuel, 2013.
[10] L. Macioszek and R. Rybski, "Low-frequency dielectric spectroscopy approach to water content in winter premium diesel fuel assessment", In: Progress Reports on Impedance Spectroscopy: Measurements, Modeling, and Application, 2016.
[11] J. C. Jones, "Hydrocarbons - Physical Properties and their Relevance to Utilisation", 2010.
[12] A. S. Bondarenko and G. A. Ragoisha, "Progress in Chemometrics Research", 2005.
[13] A. Setayeshmehr, I. Fofana, C. Eichler, A. Akbari, H. Borsi, and E Gockenbach, "Dielectric spectroscopic measurements on transformer oil-paper insulation under controlled laboratory conditions", Dielectrics and Electrical Insulation, IEEE Transactions, 2008.
[14] H. Ming and L. Toh-Ming, "Metal–Dielectric Diffusion Processes: Fundamentals", 2012.
[15] L. Macioszek, "Temperature influence on parameters of summer diesel fuel measured with the use of impedance spectroscopy", Przegląd Elektrotechniczny, 2016.

Anup Barai, Andrew McGorden, and Paul Jennings

A study of the effect of short-term relaxation on the EIS test technique for EV battery cells

Abstract: Electrochemical impedance spectroscopy (EIS) is a powerful tool employed to investigate the fundamental electrochemical parameters of a Li-ion battery cell. Its effectiveness has established it as a core method to study electrochemical behavior of batteries in laboratory environments. An industry-wide interest currently exists to employ EIS as part of battery management systems (BMSs) for state of charge, internal temperature and state of health measurements. The major issue of employing EIS for online measurement as part of BMS is the reliability. In this work, it is shown that not all the parameters extracted by the EIS technique are reliable when measurement is performed shortly after charge/discharge. While pure Ohmic resistance was found to be consistent, the change transfer resistance and double layer capacitance are not reproducible. Therefore, only R_0 can be reliably employed for the fast online measurement by BMS.

Keywords: electrochemical impedance spectroscopy, battery cells

1 Introduction

Intrinsic properties of lithium-ion (Li-ion) batteries, such as high energy, high power density, long cycle life and low self-discharge rate, have enabled their use in electric vehicle (EV) applications. Equivalent circuit models (ECMs) are commonly used by the battery management system (BMS) of an EV to estimate battery performance in real operating scenarios. In addition, investigation of the electrochemical processes, i. e., Li-ion diffusion in the electrolyte, migration through the solid electrolyte interface (SEI) layer, charge transfer through the electrode/electrolyte interface and solid state diffusion in the bulk of the active material can be achieved by using electrochemical impedance spectroscopy (EIS). As each process has a different time constant, time domain or frequency of excitation [1, 2], EIS allows for the separation of most of these processes [3]. Therefore, EIS always interests academic and industrial researchers as a diagnostic and research tool. A standard EIS test within a frequency range from 10 kHz to 10 mHz and 10 frequency points per decade of frequency, as commonly employed in the literature, requires around 10 minutes. However, if the number of frequency points is reduced, the test duration can be aggressively reduced to seconds. This potential reduction of test duration has allowed EIS to be incorporated as an integral part

Anup Barai, Andrew McGorden, Paul Jennings, WMG, University of Warwick, Coventry, CV4 7AL, United Kingdom, e-mail: A.Barai@warwick.ac.uk

https://doi.org/10.1515/9783110558920-004

of BMS to estimate the state of charge (SoC) and state of health (SoH) in EVs [4–6]. For such onboard EIS measurement systems (incorporated as part of the BMS), a short measurement time is a key requirement. However, fast measurement implies that the cells will be measured very shortly after cycling. Recently the cell relaxation period has been identified as a source of impedance variation [7, 8]. When the charge or discharge process of the cell is stopped, the polarization of the cell ceases and as a result, the cell impedance and voltage of the cell change, which is referred to as relaxation of the cell and which is mainly caused by relaxation of ions within the double layer, electrolyte ionic concentration gradient redistribution and solid state diffusion of lithium atoms within the bulk of the materials. This lead to change of cell impedance in a couple of hours after the charge/discharge event. In [8] it has been shown that a measurable change of impedance will continue until 4 hours after a discharge event. This research indicates potential hurdles to get repeatable and reproducible results from EIS tests when employed shortly after cycling. Currently there is no research present in the literature showing how reliable and reproducible the EIS tests are when performed immediately after charge/discharge. This research focuses on the impedance variation of the cell within a short period (0–10 min) after removing an electrical load from the battery cell. In Section 2 the experimental procedure is outlined. Results and discussion of the results are presented in Section 3. Finally, the findings of this study are summarized in Section 4.

2 Experimental methods

Five commercially available Li-ion cells of three different chemical natures were used as part of this study. These cells were selected based on their power and energy rating. Table 1 summarizes the chemistry, capacity and format of each cell. These cells, representing different groups of Li-ion batteries, i. e., chemistry, power capability and form factors, are currently used in different automotive applications.

In order to isolate the effects of cell relaxation, the temperature and SoC were kept constant for each test; tests were carried out at 25°C and 50 % SoC. To precisely adjust the cell SoC to 50 %, the cells were discharged at a 1C rate to the manufacturer's

Table 1: Li-ion cell electrical properties.

Cell number	Chemistry, cathode/anode	Capacity (Ah)	Format	Cell type
1	LiNiMnCoO2(NCM)/LiC6	40	Pouch	Power cell
2	LiFePO4(LFP)/LiC6	20	Pouch	Power cell
3	NCM/LiC6	2.2	Cylindrical	Energy cell
4	NCM/Li4Ti4O12(LTO)	13	Pouch	Power cell
5	NCM/LiC6	3.4	Cylindrical	Energy cell

recommended cut-off voltage, subsequently allowed to rest for 3 hours before being fully recharged according to the manufacturer's recommended charge protocol, using a commercial Li-ion cell cycler. At the end of charging, the cells were allowed to rest for 3 hours prior to being discharged to 50 % SoC at a 1C rate.

To study the short-term effect of relaxation, measurements were performed in galvanostatic mode using a commercial EIS system. Impedance measurements between 500 mHz and 10 kHz with five frequency points per decade were taken. This frequency range and this number of measurement points were selected as a balance of test duration and number of measurement points. In this setup a measurement will take around 10 seconds. Recording of the EIS data started as soon as the SoC adjustment was completed (i. e., immediately after the 1C discharge to 50 % SoC), and every minute thereafter.

3 Results

The shape of the Nyquist plots at high frequency is similar to that found in the literature. However, at low frequency, cells 1, 2 and 4 show a spiral shape (Fig. 1 (a), (b) and (d)), which has not been reported in previous related literature. The spiral shape in the Nyquist plot starts to shrink from the second test done at 1 min after SoC adjustment, and finally disappears 10 min after SoC adjustment when the EIS plot is similar to the Nyquist plots found in the literature [5–8]. The change of shape of the Nyquist plots from 1 min to 10 min for cells 1, 2 and 4 happens at a different rate. Cell 3 and 5 (Fig. 1 (c) and (e)) show this spinal shape only for the very first test, which was performed just after the discharge was stopped. Afterward the shape is similar to the Nyquist plots found in the literature. To investigate the experimental setup as a probable cause of the shape, the experiment has been repeated with different conductors, which connect the EIS equipment with the test cell, and even with different sets of EIS equipment; however, in all the instances similar results were found. The spiral shape at the low-frequency end could be related to the measurement time. For a stable EIS measurement, it is essential that there is no change of the cell during measurement at a particular frequency. At the lowest frequency employed (500 mHz), it takes 2 seconds to complete the measurement. After discharge was stopped, polarization of the double layer was ceased and as a result cell capacitance and thus total impedance change; this change slows down with the relaxation period. After 10 min of relaxation the change in the 2-second timescale is probably not significant; however, immediately after discharge it probably is significant, which leads to the spiral shape. With the difference of chemistry, size and form factor, the impedance and frequency response of each cell is different, which could be the root cause of the difference of the spiral shapes. However, for all five cells measured, values of pure Ohmic resistance (R_0) fall

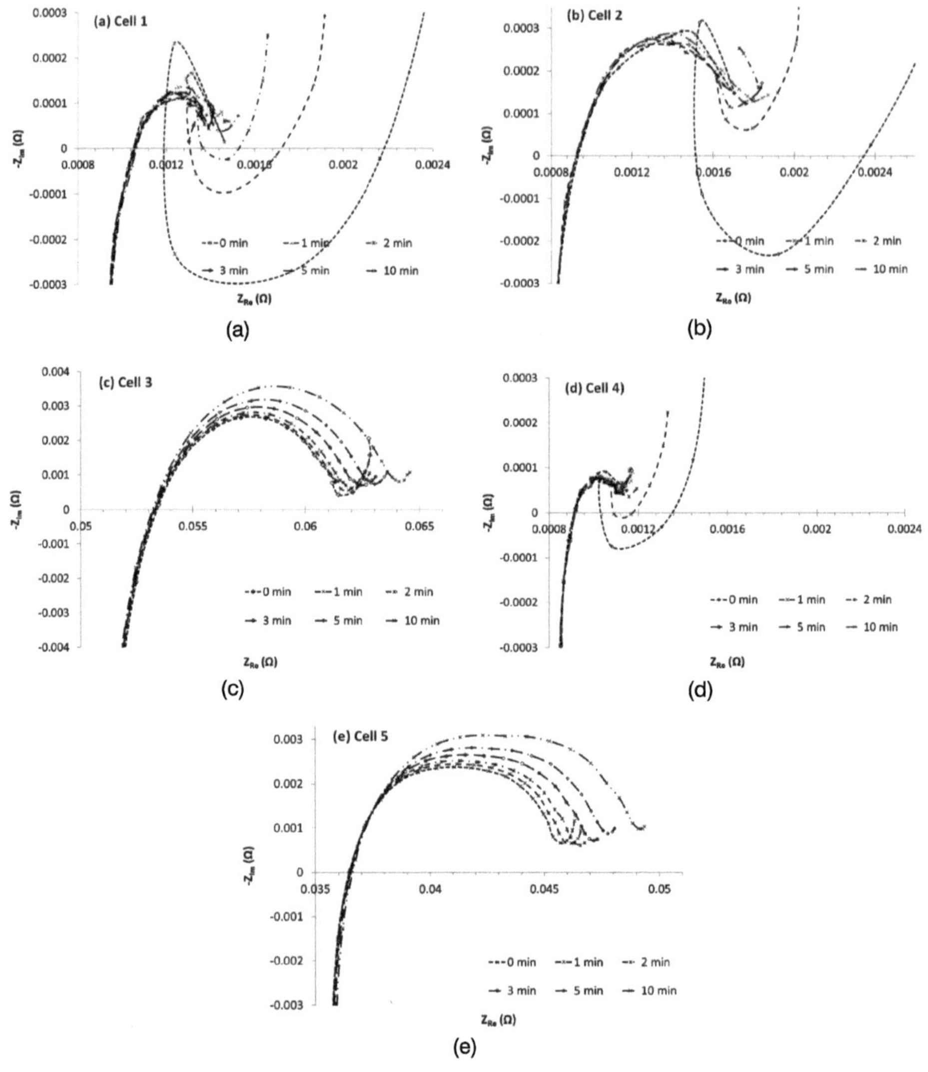

Fig. 1: Nyquist plots of different cells from 0–10 min.

within experimental error bounds, and therefore they can be considered constant, regardless of relaxation time. This is because R_0 was found at a frequency higher than 1 kHz for all the cells. Therefore, the measurement time was 1 ms or shorter and the change of the cell impedance within that short period was not significant enough to affect the measurement. Therefore, R_0 could be used as online measurement parameter. In contrast to R_0, the charge transfer resistance and double layer capacitance will be inconsistent within 0–10 minutes after cycling, thus should not be employed for fast measurement.

4 Conclusion

In this research, the variation of cell impedance as a function of relaxation period after removing an electrical load has been investigated for five different commercially available Li-ion cells. It is concluded that the pure Ohmic resistance R_0 of the cell is independent of the relaxation processes occurring within the cell for the test duration. Therefore, R_0 could be used for direct comparisons, irrespective of the relaxation period used for the test. The charge transfer resistance and double layer capacitance were found to be inconsistent within 0–10 minutes of relaxation. Therefore, when these two parameters are used as a fast online measurement parameter, they will produce unreliable results, which could affect SoC and SoH estimation. This new knowledge shows the possible inconsistency of SoC and SoH results via an EIS measurement system integrated as part of BMS, which has been proposed by researchers previously.

Bibliography

[1] E. Barsoukov and J. R. Macdonald, "Impedance Spectroscopy, Theory, Experiment, and Applications", new Jersey: John Wiley & Sons, 2005.
[2] T. Momma, M. Matsunaga, D. Mukoyama, and T. Osaka, "AC impedance analysis of lithium ion battery under temperature control", Journal of Power Sources, 2012.
[3] J. P. Schmidt, T. Chrobak, M. Ender, J. Illig, D. Klotz, and E. Ivers-Tiffée, "Studies on LiFePO4 as cathode material using impedance spectroscopy", Journal of Power Sources, 2011.
[4] S. Rodrigues, N. Munichandraiah, and A. K. Shukla, "AC impedance and state-of-charge analysis of a sealed lithium-ion rechargeable battery", Journal of Solid State Electrochemistry, 1999.
[5] A. Eddahech, O. Briat, N. Bertrand, J. Y. Delétage, and J.-M. Vinassa, "Behavior and state-of-health monitoring of Li-ion batteries using impedance spectroscopy and recurrent neural networks", International Journal of Electrical Power & Energy Systems, 2012.
[6] V. Sauvant-Moynot, J. Bernard, R. Mingant, A. Delaille, F. Mattera, S. Mailley, J.-L. Hognon, and F. Huet, "ALIDISSI, a Research Program to Evaluate Electrochemical Impedance Spectroscopy as a SoC and SoH Diagnosis Tool for Li-ion Batteries", Oil Gas Sci. Technol-Rev.IFP, 2010.
[7] F. Kindermann, A. Noel, P. Keil, and A. Jossen. "Influence of relaxation time on EIS measurements of Li-ion batteries", In: International Workshop on Impedance Spectroscopy, 2013.
[8] A. Barai, G. H. Chouchelamane, Y. Guo, A. McGordon, and P. Jennings, "A study on the impact of lithium-ion cell relaxation on electrochemical impedance spectroscopy", Journal of Power Sources, 2015.

Ammar Al-Hamry, Renato Veiga de Torres, Christian Müller, and Olfa Kanoun

Humidity sensitivity investigation of reduced graphene oxide by impedance spectroscopy

Abstract: In this chapter we report about graphene oxide (GO) and reduced GO (rGO) humidity sensitivity. The GO was reduced by thermal annealing to form rGO, which has increased its conductivity. The effects of the reduction temperature were investigated for both the electrical and humidity sensitivity properties. The electrical properties were studied by electrochemical impedance spectroscopy, which is in particular of high importance to reveal the conduction mechanism of humidity sensitivity. In addition, the morphology and optical properties of rGO films were investigated by atomic force microscopy and UV-Vis-NIR spectroscopy. The reduction temperature plays an important rule for the rGO humidity sensitivity behavior not only by change in film conductivity but also by the transport mechanism.

Keywords: graphene oxide, reduced graphene oxide, humidity sensor, impedance modeling

1 Introduction

Graphene and graphene oxide (GO) as carbon-based material with their well-known, outstanding properties have attracted a lot of attention in diverse fields, such as electronic devices like transistors and diodes [1], energy storage [2] and sensors [3, 4]. GO is considered as a very promising material due to its water dispersibility enabling solution processing [5], as well as large-scale production and easy processing. GO was used as humidity sensor either without further processing or by introducing it into a polymer matrix as well as by altering it using chemical functional groups by reduction to obtain reduced GO (rGO) [6, 7]. Reduction of GO can be carried out by several techniques, e. g., thermal, chemical or photonic reductions, by which some of the oxygen containing functional groups are eventually removed partially [8]. The importance of reduction is not only that it converts GO from a semi-insulating to a conductive material but also that it is controllable and tunable. This has a big influence on the GO and rGO electrical and sensing properties where they will be functions of type and degree of the reductions.

GO as a promising material for humidity sensors as well as other types of sensors still needs to be characterized better to explore its tunability in different aspects. The

Ammar Al-Hamry, Renato Veiga de Torres, Christian Müller, Olfa Kanoun, Chair for Measurements and Sensor Technology, Chemnitz University of Technology, Reichenhainer Strasse 70, 09126 Chemnitz, Germany, e-mail: ammar.al-hamry@s2011.tu-chemnitz.de

https://doi.org/10.1515/9783110558920-005

ability to lose their oxygen groups by reduction and to show different surface, conductivity and sensitivity properties is of high interest. In this chapter, we investigate the effect of the reduction temperature on the humidity sensitivity of GO films. The measurements were carried out by taking the impedance of rGO films.

2 Experimental procedures

GO-coated thin films were prepared by spin coating. The reduction process was carried out by thermal annealing at different temperatures to produce rGO. Electrical contacts were made by silver paste. The humidity measurements were taken under certain relative humidity (RH) conditions by means of saturated slat solution at RH values of 0 %, 11.3 %, 32.5 %, 52.3 %, 75.2 %, 84.3 % and 97.3 % and at room temperature (23°C). Electrical impedance of the films was measured by the Agilent impedance analyzer 4294A.

3 Results and discussion

In Fig. 1, the impedance relation to the humidity of rGO films reduced at different temperatures is shown. The change in impedance by thermal heating is one order of magnitude by increasing heat from 160°C to 250°C, where GO itself is semi-insulating, having a resistance of hundreds of MOhms. The sensitivity behavior shows dependency on the reduction temperature. At low humidity, the impedance increases for all rGO films. At high humidity, the impedance tends to decrease and shows a negative slope. However, for the films reduced at higher temperature, the impedance shows a positive slope. The relation is quite linear at lower humidity values.

In the Nyquist plot, Fig. 2 (a) and (b), a semicircle fits the parallel resistance and capacitance, which represent the intrinsic impedance of the film. The behavior at low

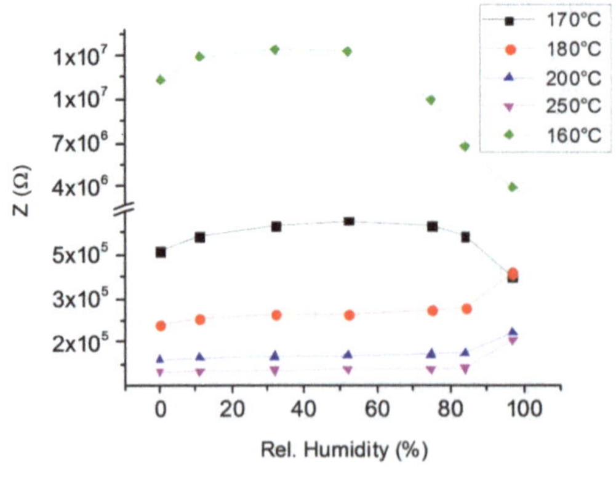

Fig. 1: Relation of impedance of rGO films with RH at 100 Hz.

humidity range (0–52 %) shows an increase of the semicircle due to absorbed water molecules. This causes to charge transfer between water molecules and rGO, which reduces the charge carriers of the film [7]. This is applied for all the films. At high humidity (75–97 %), in the Nyquist plot of Fig. 2 (b), the semicircle with small tail shows a constant phase element (CPE) behavior. The equivalent circuit consists of two CPEs and a resistance that represents the interaction of water and contact and intrinsic resistance.

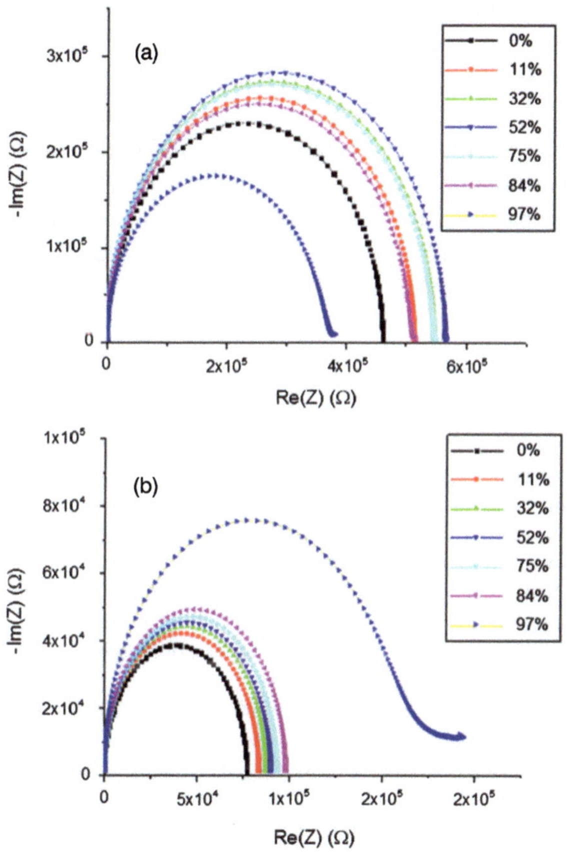

Fig. 2: Nyquist plot for (a) rGo reduced at 170 °C and (b) rGO reduced at 200 °C under different RH conditions.

For rGO reduced at higher temperatures, by increase in the number of water molecules, the charge transfer is dominant and causes a further increase of the impedance. For low-temperature reduced films, the formation of water thin films has the role of decreasing the impedance due to proton hopping [9]. In addition, by increasing the reduction temperature, rGO films become more hydrophobic, which influences the formation of a continuous water film. Thus the impedance only further increases at higher values of humidity.

4 Summary and conclusion

GO coated and reduced by temperature leads to tuning not only the conductivity but also the film sensitivity to humidity. The impedance can be reduced from hundreds of MOhms to tens of kOhms by heating the films to 250°C. The reduction by temperature results in different humidity sensitivity behaviors. Tuning the reduction temperature provides useful selectivity of the humidity sensitivity. Where at lower temperature is more sensitive at higher humidities, at high reduction temperatures, the sensitivity to humidity becomes less due to the increase of the hydrophobicity of the rGO films. However, it improves the linearity.

The realization of a humidity sensor or sensor matrix that covers the full range of humidity is also foreseen, e. g., by combining two films with two different annealing temperatures and thus different transport mechanisms at low and high temperature. The electrical impedance was obtained by impedance spectroscopy, which reveals the conduction mechanism. However, further tests and study of the equivalent circuit models will be a next step, which will provide an in-depth understanding of the different behaviors in humidity sensitivity of rGO.

Bibliography

[1] J. Basu, J. K. Basu, and T. K. Bhattacharyya, "The evolution of graphene-based electronic devices", International Journal of Smart and Nano Materials, pp. 201–223, 2010.
[2] R. Mukherjee, A. V. Thomas, D. Datta, E. Singh, O. Eksik et al., "Defect-induced plating of lithium metal within porous graphene networks", Nature Communications, 2014.
[3] Z. Bo, X. Shuai, S. Mao, H. Yang, J. Qian, J. Chen et al., "Green preparation of reduced graphene oxide for sensing and energy storage applications", Scientific Reports, 2014.
[4] O. Kanoun, C. Müller, A. Benchirouf, A. Sanli, T. Dinh, A. Al-Hamry et al., "Flexible carbon nanotube films for high performance strain sensors", Sensors, 2014.
[5] G. Eda and M. Chhowalla, "Chemically derived graphene oxide: towards large-area thin-film electronics and optoelectronics", Advanced Materials, 2010.
[6] S. Borini, R. White, Di. Wei, M. Astley, S. Haque, E. Spigone et al., "Ultrafast graphene oxide humidity sensors", ACS Nano, vol. 7, 2013.
[7] D. Zhang, J. Tong, and B. Xia, "Humidity-sensing properties of chemically reduced graphene oxide/polymer nanocomposite film sensor based on layer-by-layer nano self-assembly", Sensors and Actuators B, Chemical, vol. 197, 2014.
[8] A. Al-Hamry, H. Kang, E. Sowade, V. Dzhagan, R. Rodriguez, C. Müller, D. Zahn, R. Baumann, and O. Kanoun, "Tuning the reduction and conductivity of solution-processed graphene oxide by intense pulsed light", Journal of Carbon, vol. 102, 2016.
[9] D. Zhang, J. Tong, B. Xia, and Q. Xue, "Ultrahigh performance humidity sensor based on layer-by-layer self-assembly of graphene oxide/polyelectrolyte nanocomposite film", Sensors and Actuators B, Chemical, vol. 203, 2014.

Noureddine Leguedani, My Mustapha Hafid,
Siham Yazami el Idrissi, and Yousra kessad

Impedance study of Ni-Cr alloy in contact with biologic and organic solution using constant phase element

Abstract: The electrochemical characterization of the Ni-Cr dental implant alloy was studied by electrochemical impedance spectroscopy (EIS) in contact with artificial saliva, milk and vinegar. For all solutions the data impedance has been recorded maintaining the sample at 37°C for 1 hour of immersion. In this chapter we are focused on the modeling of the response function of this biomaterial using an equivalent electric circuit (EEC). The experimental results show that the Ni-Cr alloy exhibits a quite good resistance to corrosion, with passive properties referring to the existence of a thick passive film. An EEC was proposed to describe the electrochemical behavior of this biomaterial in the frequency range of 0–40 kHz. The Bode diagram indicates the presence of at least one time constant during the kinetic process. The nature of the film and the morphology of the interface biomaterial/film/electrolyte are presented in this chapter. The proposed circuit involved constant phase elements in order to explain the apparent heterogeneity on the interface biomaterial/film/electrolyte. Simulation based on this model showed a good agreement with the experimental data, with a best fit corresponding to a minimum standard deviation (the chi-square value χ^2 is 10^{-4}).

Keywords: Ni-Cr, biomaterial, electrochemical impedance spectroscopy (EIS), electrical equivalent circuit (EEC), constant phase element (CPE), interface biomaterial/film/electrolyte

1 Introduction

Ni-Cr-based alloys are part of materials used on orthodontics as dental prostheses. Their study has become easier owing to their lower cost, by their biocompatibility and by the existence of a variety of software able to fit the data impedance. The Ni-Cr alloys are commonly used for dental materials [1] and widely used in dental skeletal structures, implant fixtures and prosthodontic restoration [2]. Beside their low cost they exhibit acceptable mechanical and tribological properties and matching thermal expansion coefficient with the ceramics of metal–ceramic restorations [3, 4]. There also known as passive system with a spontaneous formation of thin passive oxide film

Noureddine Leguedani, My Mustapha Hafid, Siham Yazami el Idrissi, Yousra kessad, Laboratory of Engineering Physics and Environment, Department of Physics, Faculty of Science, University Ibn Tofail, Kenitra, Morocco, e-mail: leguedani@yahoo.fr

https://doi.org/10.1515/9783110558920-006

on the interface metal/electrolyte [5]. The reactivity of the oxide superficial layer formation, which is based especially on Cr_2O_3, will be investigated by electrochemical impedance spectroscopy (EIS). Moreover, we study the reactivity of the superficial oxide layer while varying the solution (pH). We used three solutions, namely, artificial saliva (pH $\approx$ 6.8), milk (pH $\approx$ 6.5 to 6.7) and vinegar (pH $\approx$ 3), to evidence the reactivity of the interface. In this work, we compare the Ni-based alloys to the Co-based alloys, with their formed protective film depending highly on the acidity of the solution used [6]. We study by EIS the response of the biomaterials and the reactivity of its passive film, formed with equivalent electric circuit (EEC) involving CPE.

2 Electrochemical characterization

2.1 Impedance spectroscopy measurements

The electrochemical impedance spectroscopy, or also AC impedance, is a non-destructive technique and a useful tool for studying corrosion [7]. This experimental method makes it possible to distinguish the dielectric and electric properties of individual contributions of components [5].

In this method, the sample undergoes a small electronic excitation so its response is pseudo-linear. In a linear (or pseudo-linear) system, the current response (1) and the potential (2) are sinusoidal in the same frequency and shifted in phase.

The sinusoidal perturbation of potential $E(t)$ induces a sinusoidal current $I(t)$; their representations in the polar coordinates is as follows:

$$E(t) = E_0 \sin(\omega t), \tag{1}$$

$$I(t) = I_0 \sin(\omega t + \varphi), \tag{2}$$

where $\omega = 2\pi f$ and E_0 and I_0 are the amplitudes of the potential and the current, respectively.

The impedance of the system (3) can be written following the Ohmic concept as follows:

$$Z(t) = \frac{E(t)}{I(t)} = \frac{E_0 \sin(\omega t)}{I_0 \sin(\omega t + \varphi)} = Z_0 \frac{\sin(\omega t)}{\sin(\omega t + \varphi)}. \tag{3}$$

With Euler's relationship or polar representation of $E(t)$ and $I(t)$, they can be written as follows:

$$E(t) = E_0 e^{(j\omega t)}, \tag{4}$$

$$I(t) = I_0 e^{(\omega t - \varphi)}. \tag{5}$$

The impedance complex (6) of the biomaterials can be rewritten as follows:

$$Z(t) = \frac{E(t)}{I(t)} = Z_0 e^{j\varphi} = Z_0(\cos\varphi + \sin\varphi). \tag{6}$$

The modulus of the complex impedance (7) and the phase angle shift (8) are deduced from the following formulas:

$$|Z(\omega)| = \sqrt{Z_{\text{re}}^2 + Z_{\text{im}}^2}, \tag{7}$$

$$\tan(\varphi) = \frac{Z_{\text{im}}}{Z_{\text{re}}}, \tag{8}$$

where Z_{re} and Z_{im} are the real and imaginary part of the complex impedance, respectively.

The impedance data are profoundly interpreted by much recently developed software. We have used ZSimpwin version 3.10 to fit the response function of the Ni-Cr biomaterial. The EEC consisted of a combination of simple electronic elements (resistance, capacitor and inductance). The numerical simulation must be done with more accuracy to solve difficulties related to the fact that more than one EEC can be able to fit the data.

Studying the response function $Z(\omega)$ we return to interpret two famous diagrams or representations, namely, the diagrams of Bode and Nyquist. There is some literature which talks about the basics of EIS [8–13], and the fields of its applications cover many domains, from biology and medicine to industry. The electrochemical impedance analysis has been performed on the commercial Ni-Cr alloy in the frequency range of 0–40 kHz, after it had been immersed for 1 hour in milk, artificial saliva or vinegar.

2.2 Experimental results

The Bode diagrams (Fig. 1 and Fig. 2) of the Ni-Cr dental alloys recorded in milk, artificial saliva and vinegar show that at least one time constant phase element (CPE) can be distinguished. For artificial saliva and vinegar, the shape of the curves are similar. At high frequencies the absolute impedance curve is almost independent of the frequency, with a phase angle of 0°, and the impedance of the system is reduced to the electrolyte resistance.

For medium frequencies, at 1–100 Hz, a linear frequency dependence of absolute complex impedance with a maximum angle reaching 70°, 65° and 55° for milk, artificial saliva and vinegar, respectively. This range of frequency is marked by the stable film formation.

At low frequencies, the absolute impedance exhibits its high values for milk and artificial saliva, and these values are ten times higher than for vinegar. This difference in the values of the impedance gives an idea of the film formation for the three

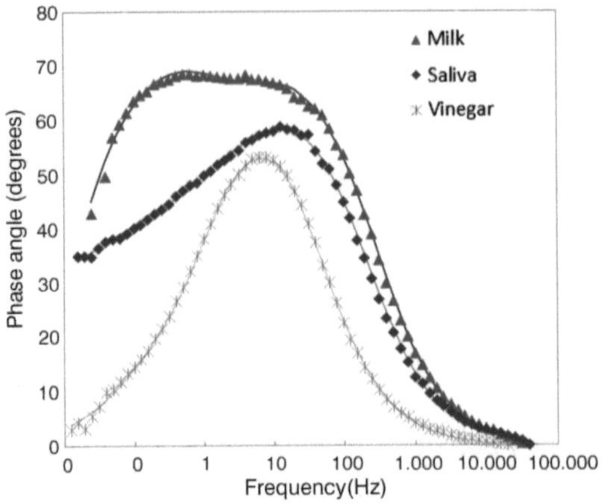

Fig. 1: Phase angle plot for Ni-Cr for milk, saliva and vinegar. The fit based on the model is represented by solid lines.

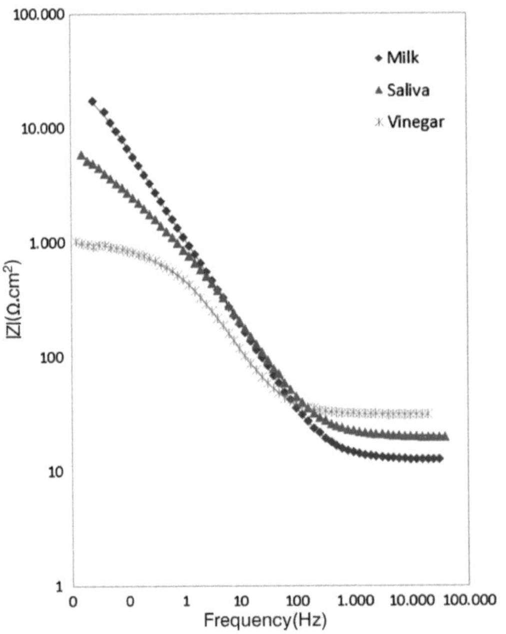

Fig. 2: Measured (discrete points) and fitted (solid lines) values of the Z modulus for Ni-Cr dental alloys for milk, saliva and vinegar for 1 hour immersion at 37°C.

solutions under study. This can be explained by the fact that the film is of a porous character in origin. x

The Nyquist plots of Ni-Cr dental alloys are shown in Fig. 3 for the three solutions (milk, artificial saliva and vinegar) with the potential corrosion E_{corr} and at 37°C. It can be seen from the impedance data that the Ni-Cr implant biomaterial/solution interface exhibits capacitive behavior over a relatively wide frequency region, which is typical for a passive alloy system. The parameters for the solution bulk resistance R_s (experimentally determined by the intersection at high frequency of the Nyquist plot

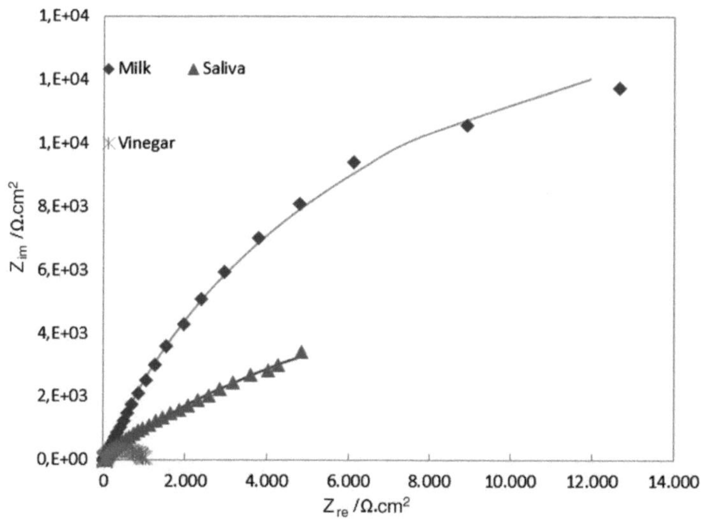

Fig. 3: Nyquist diagram of Ni-Cr dental alloys in the three solutions of milk, artificial saliva and vinegar for 1 hour immersion at 37°C (discrete points).

with the real axis) are of about the same order of magnitude in the three solutions; 12.63 Ω.cm^2 for milk, 19.88 Ω.cm^2 for artificial saliva and 30.9 Ω.cm^2 for vinegar.

The resistance of polarization, R_p (experimentally determined by the diameter of the semi-circles on the arc in the Nyquist plot), related to the corrosion resistance, is found to have high values that increase significantly with milk, artificial saliva and vinegar, respectively.

This high value of R_p implies higher corrosion resistance of implant alloy for milk, reaching a value of 25000 Ω.cm^2, while this value is about 1600 Ω.cm^2 for artificial saliva and 1000 Ω.cm^2 for vinegar.

The low R_p values in vinegar show that the Ni-Cr dental alloys have a weak corrosion resistance when compared to milk and artificial saliva; then a high rate of released metallic ions into the oral environment can take place. The Cole-Cole plot of the complex admittance (Fig. 4) shows the same geometry and the arc dependence of the semi-circle is affected by the nature of the solution used. The diameter of the semi-circle decreases with vinegar, artificial saliva and milk, respectively.

2.3 Modeling the impedance data

The proposed equivalent circuit based on CPE is described in Fig. 5, where R_s is the electrolyte resistance, in series with Q_{out} and in parallel with R_{out}, which is in series with Q_{in}, which is in parallel with R_{in}. Here, Q_{out} denotes the pseudo-capacitance (CPE_{out}) and R_{out} the resistance of the Ni-Cr protection film system, whereas Q_{in} and

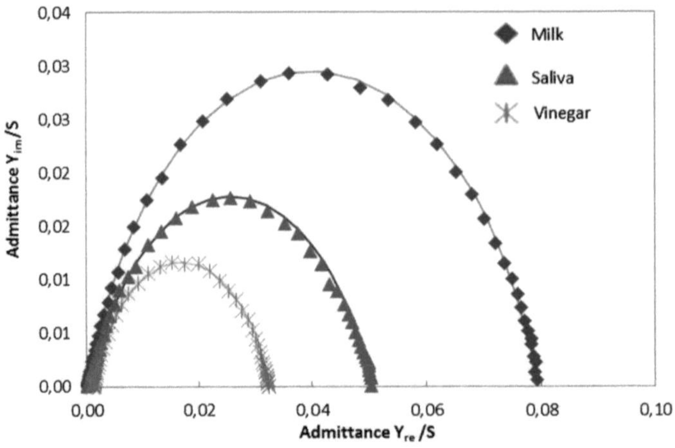

Fig. 4: Admittance diagram of Ni-Cr dental alloys in the three solutions of milk, artificial saliva and vinegar for 1 hour immersion at 37° (discrete points).

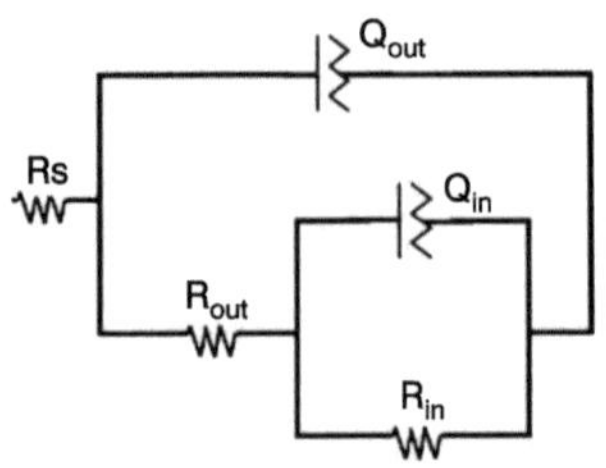

Fig. 5: EEC used in the generation of simulated data.

R_{in} denote the double layer pseudo-capacitance (CPE_{in}) and the transfer resistance of the corrosion of the Ni-Cr/solution interface.

CPE (9) was mathematically used to solve the response of the biomaterial with a non-ideal capacitor interpreting the non-homogeneous behavior of the surface under investigation. The main reasons for adding CPE in a matrix of EEC were the distributed surface reactivity, surface inhomogeneity, roughness and electrode porosity [6]. This lack of homogeneity is modeled with a Q element, used to represent the CPE, i. e.,

$$Z_Q(\omega) = \frac{1}{Y_0(j\omega)^n}, \tag{9}$$

where Y_0 is the admittance of an ideal capacitance and n is an empirical constant, ranging from 0 to 1. The exponent n is related to a non-uniform current distribution due to the surface roughness [7, 8]. It is noteworthy that when $n = 1$, the CPE behaves as a pure capacitor, while when $n = 0$, the CPE behaves as a pure resistor and when $n = 0.5$, the CPE is the equivalent of the so-called Warburg element.

The impedance of ZCPE is defined by $ZCPE_{out} = 1/(j\omega)^{n1}Q_1$ and $ZCPE_{in} = 1/(j\omega)^{n2}Q_2$, where $j^2 = -1$, ω is the angular frequency and the exponents n_1 and n_2 of the CPE_{out}

and CPE_{in} are related to a non-uniform current distribution due to the surface roughness [12]. The chi-square values (χ^2) for all data impedance are of magnitude 10^{-4} indicating an excellent agreement between the experimental data and the model using CPE in the fitting program.

3 Simulation criterion and discussion

Taking into account that several circuits can fit the same data impedance, we aim to determine which of the circuits is more suitable to best describe the corrosion phenomena. We used two criteria to decide which circuits numerically fit the impedance data in search of an optimal configuration. First, only the simple elements introduced into the EEC must have a physical meaning and second, the chi-square value (χ^2) must be less than 10^{-4}. Monitoring numerically the evolution of parameters, one can determine which of the proposed models can be kept. Systematically, the standard deviation in the present work is found to be 1.732×10^{-04} for milk, 3.895×10^{-04} for artificial saliva and 1.693×10^{-04} for vinegar.

From this model it was possible to calculate the polarization resistance, the resistance R_s of the solution, the relaxation frequency of the system and the CPE parameters (the resistance and the capacity of the oxide layer). Thus a simple schematization of the interface (Fig. 6) can be given.

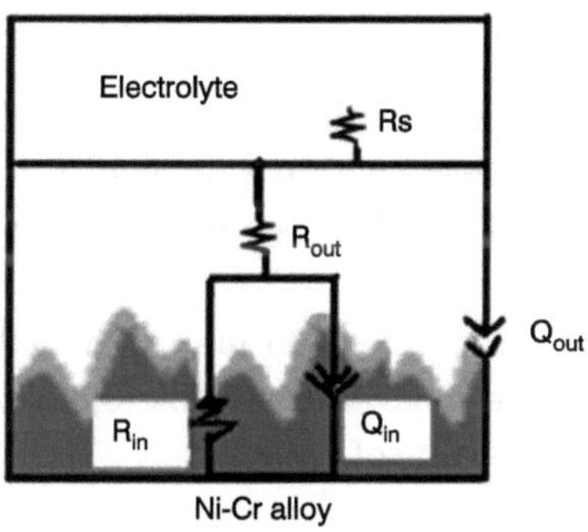

Fig. 6: Modeling of the interface electrode/film/electrolyte of Ni-Cr dental alloys.

Table 1 lists the parameters of the EEC used to simulate the impedance data of the Ni-Cr implant alloy in the presence of milk, artificial saliva and vinegar, maintained at 37°C and for one hour immersion time.

It was found that the values of the resistance of the electrolyte, R_s, for the Ni-Cr implant alloy is increasing with milk (12.63 $\Omega.cm^2$), artificial saliva (20.06 $\Omega.cm^2$) and vinegar (31.06 $\Omega.cm^2$). The resistance of polarization (Fig. 7) is calculated in this model by the sum of R_{in} and R_{out}. The value of R_p decreased; it is 33.588 $k\Omega.cm^2$ for milk, 28.579 $k\Omega.cm^2$ for saliva and 991.2 $\Omega.cm^2$ for vinegar.

Table 1: Model parameters used to simulate the impedance spectra of the Ni-Cr implant alloy in the presence of milk, artificial saliva and vinegar.

	R_s Ω.cm^2	Q_{out} μF.cm^2	n_1	R_{out} Ω.cm^2	Q_{in} μF.cm^2	n_2	R_{in} Ω.cm^2
Vinegar	31.06	313.6	0.82	715.5	3734	0.72	275.7
Milk	12.63	171.9	0.82	2848	35.5	0.89	3.07×10^{04}
Saliva	20.06	163.5	0.8	699	380.4	0.44	2.79×10^4

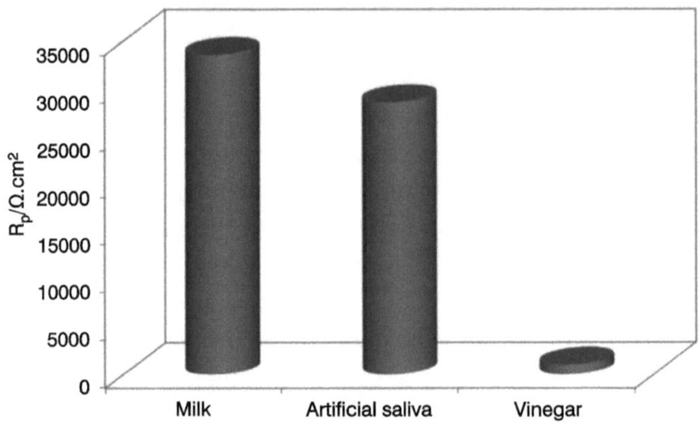

Fig. 7: Resistance of polarization, R_p, of Ni-Cr alloys in the three solutions of milk, artificial saliva and vinegar.

The theoretical model predictions are in good agreement with the experimental data. Thus, the Q_{out} capacitance values in all solutions were within a range of 172–314 μF.cm^2, with n within a range of 0.8029–0.8213, which indicates a non-ideal capacitance interface (so an oxide passive layer exists).

Moreover, the Q_{in} capacitance values are highly dependent on milk, artificial saliva and especially of vinegar; the values of n have the same order in milk (0.8924) and in vinegar (0.7151), but not in saliva, which shows a value of 0.4361.

4 Conclusion

The corrosion interface of Ni-Cr alloys in biological (milk and artificial saliva) and organic solutions (vinegar) was studied by EIS based on a model using CPE. Results show that this biomaterial exhibits a good corrosion resistance in milk and artificial saliva and a weak corrosion resistance in vinegar. All the n values in the three solutions were found to be within a range of 0.7151 to 0.8924, indicating that the corrosion interface deviated from a pure ideal capacitor. The phase angle maximum (φ_{max}) in this biomaterial is measured at approximately –53 to –70 degrees. The EEC predictions are in good agreement with the experimental data.

Bibliography

[1] K. F. Leinfelder, "An evaluation of casting alloys used for restorative procedures", The Journal of the American Dental Association, vol. 128, pp. 37–45, 1997.

[2] H.-S. Kima, M.-K. Sonb, and H.-C. Choec, Journal of the Korean Institute of Surface Engineering, vol. 49, no. 3, 2016.

[3] J. C. Wataha, N. L. O'Dell NL, B. B. Singh, M. Ghazi, G. M. Whitford, and P. E. Lockwood, "Relating nickel-induced tissue inflammation to nickel release in vivo", Journal of Biomedical Materials Research, vol. 58, pp. 537–544, 2001.

[4] R. M. Joias, R. N. Tango, J. E. Junho de Araujo, M. A. Junho de Araujo, Gde S. Ferreira Anzaloni Saavedra, T. J. Paes-Junior, and E. T. Kimpara, "Shear bond strength of a ceramic to Co-Cr alloys", The Journal of Prosthetic Dentistry, vol. 99 pp. 54–59, 2008.

[5] H. Cesiulis, N. Tsyntsaru, A. Ramanavicius, and G. Ragoisha, "Nanostructures and thin films for multifunctional applications", NanoScience and Technology, pp. 3–42, April 2016.

[6] A. Ouerd, C. Alemany-Dumont, B. Normand, and S. Szunerits, Electrochimica Acta, vol. 53, pp. 4461–4469, 2008.

[7] A. S. Hamdy, E. El-Shenawy, and T. El-Bitar, International Journal of Electrochemical Science, vol. 1, pp. 171–180, 2006.

[8] M. Sluyters-Rehbah, "Impedances of electrochemical systems: terminology, nomenclature and representation. Part 1. Cells with metal electrodes and liquid solutions". Pure and Applied Chemistry, vol. 66, no. 9, pp. 1831–1891, 1994.

[9] A. J. Bard and L. R. Faulkner, "Electrochemical Methods. Fundamentals and Applications", Wiley, New York, 2001, 829 pp.

[10] E. Barsoukov and J. R. Macdonald (eds.), "Impedance Spectroscopy Theory, Experiment, and Applications", Wiley, New York, 2005, 595 pp.

[11] M. E. Orazem and B. Tribollet, "Electrochemical Impedance Spectroscopy", Wiley, New York, 2008, 523 pp.

[12] A Lasia, In: "Electrochemical Impedance Spectroscopy and its Applications", ed. by B. E. Conway, J. O'M. Bockris, R. E. White, Modern Aspects of Electrochemistry, vol. 32, Kluwer Academic Publishers, New York, pp. 143–248.

[13] A. Lasia, "Electrochemical Impedance Spectroscopy and its Applications", Springer, New York, 2014, 367 pp.

Part II: **Measurement systems**

Thorsten Baumhöfer, Susanne Rothgang, and Dirk Uwe Sauer

Influence of the settling behavior on the measurement uncertainty of EIS and how to decrease the measurement duration

Abstract: To obtain a spectrum, different frequencies are applied stepwise. In an ideal system the measurement uncertainty is only influenced by the noise of the voltage and current measurements. But since a switched sine includes transient signal parts, the system's settling behavior leads to systematic errors. Usually the first periods after the start of the excitation are thus not included in the measurement. A method based on a model of the measurement system will be shown, allowing for compensation of the error. Thus the measurement duration can be reduced significantly.

Keywords: Uncertainty, settling behavior, measurement model

1 Introduction

The electrochemical impedance spectroscopy is widely used in laboratories as a tool for the parameterization of electrochemical energy storage models based on equivalent circuits [1]. It is also used as a tool to investigate changing processes in aging batteries [4]. A good knowledge of the measurement uncertainty is needed in both cases to be able to evaluate the determined parameters.

Another growing field of application is quality assurance during the manufacturing of battery packs. In a series production line the reduction of cycle time has top priority. Therefore, the goal is to decrease the measurement duration of an impedance spectrum, without increasing the measurement uncertainty considerably.

The most common way to measure the electrochemical impedance is to apply a sinusoidal excitation signal. The excitation and response signals are acquired using a constant sample rate. Both signals are then transferred to the frequency domain by using the discrete Fourier transformation and divided to obtain the impedance. To acquire a spectrum, different frequencies are applied stepwise, one after another.

To reduce the measurement duration, multiple frequencies can be applied simultaneously [2, 3]. Since the information density of an electrochemical system is not

Acknowledgement: This work was sponsored by the German Federal Ministry of Education and Research (BMBF) under contract no. 13N12035 as part of the project "eProduction". The sole responsibility for the content of this chapter lies with the authors.

Thorsten Baumhöfer, Susanne Rothgang, Dirk Uwe Sauer, Electrochemical Energy Conversion and Storage Systems Group, Institute for Power Electronics and Electrical Drives, RWTH Aachen University, Aachen, Germany, e-mail: Thorsten.Baumhoefer@isea.rwth-aachen.de

https://doi.org/10.1515/9783110558920-007

scaled linearly with frequency, only a finite number of distinct frequencies has to be included in the signal. The phase angles can further be chosen in a way to minimize the crest factor [5]. Overall suitable signal-to-noise ratios can be reached.

2 Measurement uncertainty

The main influence on the uncertainty of measured impedance values, which cannot be reduced by calibration, is the noise of the voltage and current measurement.

With the Fourier transformation the noise component at the examined frequency gets isolated and results in a Gaussian distribution around the true impedance value [6]. This behavior was verified by 22,000 measurements on a Sanyo/Panasonic UR18650E Li-ion cell, each with one period at 220 Hz. The resulting histogram (Fig. 1) shows the analytically proven Gaussian distribution.

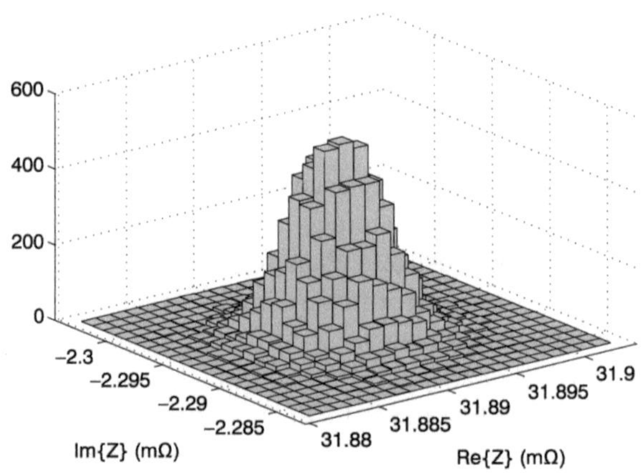

Fig. 1: Gaussian distribution of impedance measurement values due to noise in the excitation and response signals.

The standard deviation of the distribution depends on the noise magnitude at the considered frequency, the magnitude of the excitation and response signals and the number of samples used for the calculation.

3 Settling influence

Electrochemical energy storage devices exhibit processes with different time constants. This results in a transient response, when the excitation signal is being switched on. It takes a specific time until the amplitude and phase shift can be considered stable. To apply a Fourier transformation to this unstable part can be considered wrong, since the assumption of a periodic signal is not valid. But still it can be computed without numerical difficulties and provide information about the system.

To quantify the influence of the settling 20 periods at 220 Hz where applied to the previously introduced Li-ion cell. This procedure was repeated 100 times with breaks without excitation in between. Every period was individually evaluated, resulting in the impedance points seen in Fig. 2. For the first cycle the error due to the settling behavior is quite significant compared to the uncertainty due to noise. The error decreases exponentially with increased cycle number, but in this case has to be considered as a systematic error up to cycle number five.

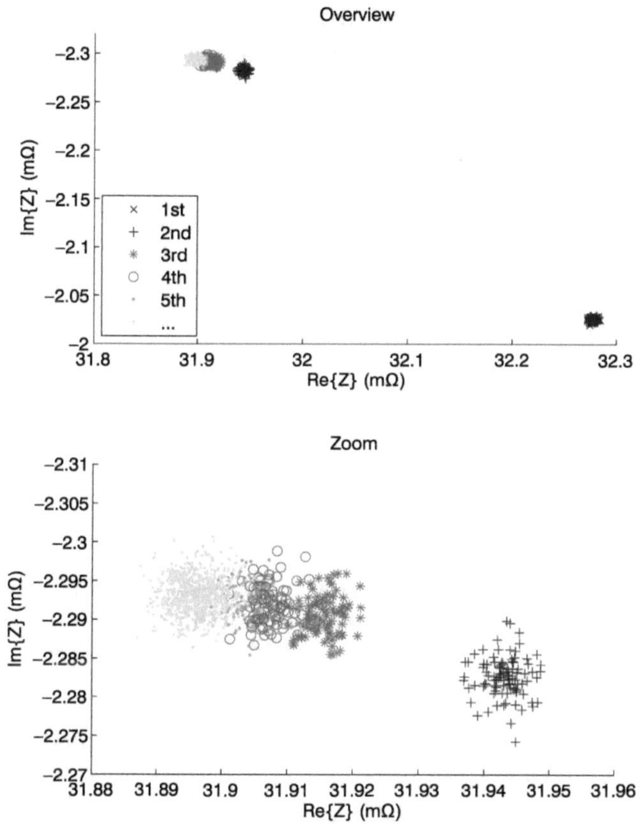

Fig. 2: Results of multiple repeated measurements on a Li-ion cell marked depending on the number of the considered excitation period.

The error is not constant but depends on the time constants and resistive magnitude of the analyzed system and changes in magnitude and direction with the frequency being applied.

The usual way to overcome this effect is to apply the first period but start the acquisition afterwards, during the second cycle. This is documented by the different instrument manufactures in different detail (e. g., as “pre-wave” [7]). Beginning from the second cycle the following cycles are usually averaged to reduce the noise level. The effect of the settling is also being reduced by this averaging.

4 Reduction of measurement duration

If the measurement duration should be reduced, long averaging cannot be used anymore. It would further be favorable to not have to apply at least two periods of the signal. Especially at lower frequencies, in the mHz range, this is considerably influencing the measurement duration.

In Fig. 3 the resulting spectrum of the first excitation period and the true system response (from the 20th period) are shown. For these measurements a multisine signal with 39 frequencies, logarithmically spaced from 8 kHz to 100 mHz, was applied to the previously introduced cell. The length of the signal is 10 seconds (one period of the lowest frequency). It can be seen that the influence in the lower frequencies is greater than in the higher frequencies. This is due to deviating numbers of periods of the different frequency components and therefore a variation in the averaging.

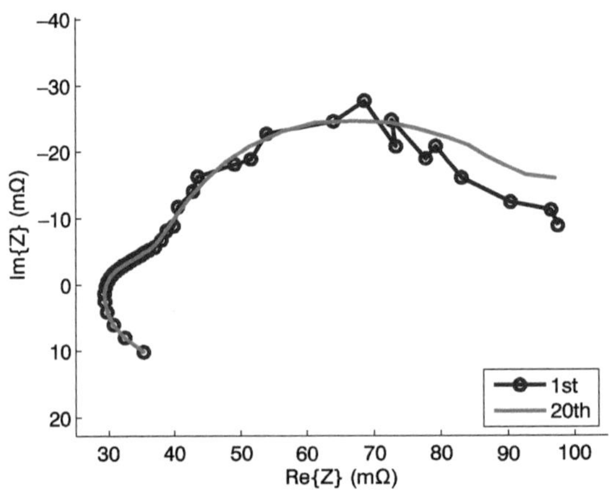

Fig. 3: Spectra measured with multisine excitation. The settling error can clearly be identified within the first period.

In the measurement model (see Fig. 4) the signal source, amplifier and input filter are assumed to be ideal. The multisine signal is periodic, so the gating in the current measurement path has no effect. But the gating of the voltage signal is the source of the settling error and thus has to be considered in the model.

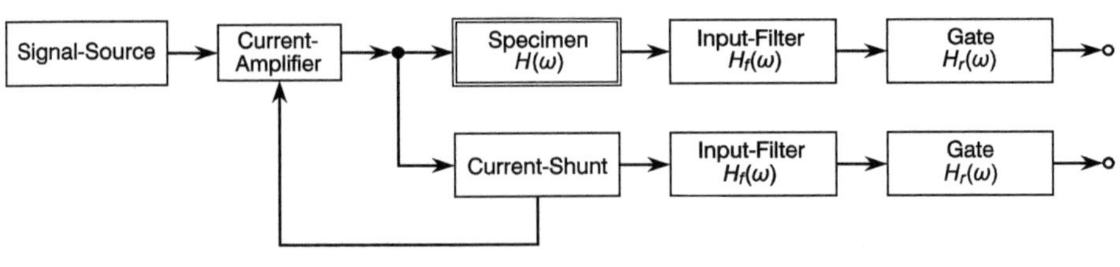

Fig. 4: Model of the measurement system.

Overall, equation (1) can be derived from the model, where $H(\omega)$ denotes the true impedance of the system, $H_r\omega$ is the transfer function of an ideal rectangular window and V, I can be observed. A response impedance $H^*(\omega)$ can be calculated, including the settling error. The equation cannot be de-convoluted to be solved with respect to H, since the problem is ill defined and would lead to massive noise amplification. We have

$$V(\omega) = \big(I(\omega) \cdot H(\omega)\big) \star H_r(\omega), \tag{1}$$

$$H^*(\omega) = \frac{V(\omega)}{I(\omega)}. \tag{2}$$

To reconstruct the true impedance, an iterative method has to be developed only computing the measurement model in forward direction.

1. Set the system impedance input $H_{n=0}$ to the measured response impedance H^* interpolating in between the measured frequencies.
2. Calculate the new response impedance H_n^* with the measurement model.
3. Change the input by the difference of the responses $H_{n+1} = H_n + (H_n^* - H_{n-1}^*)$ at the frequency points excited. Apply an interpolation as above.
4. Repeat from step 2 until the iteration converges.

An interpolation can be used without introducing errors, since electrochemical energy storage devices exhibit continuous impedance functions not including discontinuities.

The first three iterations on the previously shown measurement results are shown in Fig. 5. It can be seen that the curve of the input impedance starts to smoothen and the response impedance converges towards the measured curve. It is therefore possible to reconstruct the true impedance of the system only taking the measurement values from the first period into account.

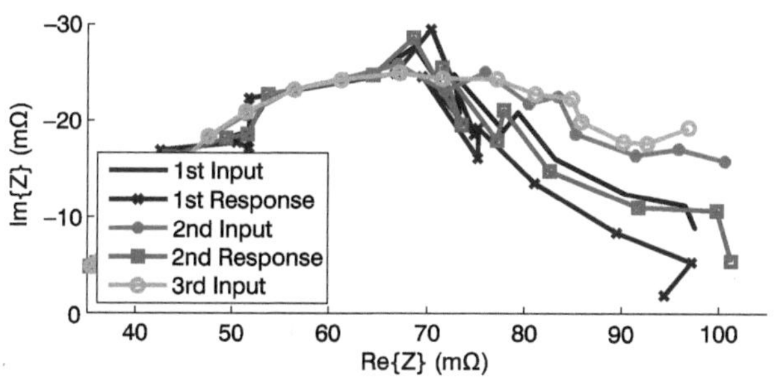

Fig. 5: Presentation of the first three iterations of the proposed method to reconstruct the true impedance.

5 Conclusion

The settling behavior is a significant error source for impedance determination within the first to sixth excitation period. A method was given to reconstruct an impedance spectrum utilizing a measurement model. It is sufficient to excite the system with a multisine excitation with the length of one period of the lowest frequency. The measurement duration for impedance spectra can thus be significantly reduced.

Bibliography

[1] E. Barsoukov and J R. Macdonalds, "Impedance Spectroscopy: Theory, Experiment & Applications", Wiley, 2005.
[2] J. P. Christophersen, W. H. Morrison, J. L. Morrison, C. G. Motloch, and D. M. Rose, "Crosstalk compensation for a rapid, higher-resolution impedance spectrum measurement", In: IEEE Aerospace Conference, pp. 1–16, 2012.
[3] J. P. Christophersen, J. Morrison, W. Morrison, and C. Motloch, "Rapid impedance spectrum measurements for state-of-health assessment of energy storage devices", SAE International Journal of Passenger Cars–Electronic and Electrical Systems, 2012.
[4] M. Ecker, N. Nerea, S. Käbitz, J. Schmalstieg, H. Blanke, A. Warnecke, and D. U. Sauer, "Calendar and cycle life study of Li(NiMnCo)O2-based 18650 lithium-ion batteries", Journal of Power Sources, pp. 839–851, 2014.
[5] Gamry Instruments, Inc, "OptiEIS: A Multisine Implementation Application Note", 2011.
[6] M. A. Richards, "The Discrete-Time Fourier Transform and Discrete Fourier Transform of Windowed Stationary White Noise", 2007.
[7] ZAHNER-Elektrik GmbH & Co.KG, ZENNIUM Operation Manual, 2014.

Achraf Lamlih, Vincent Kerzérho, Serge Bernard, Fabien Soulier, Mariane Comte, Michel Renovell, Tristan Rouyer, and Sylvain Bonhommeau

Mixed-level simulation tool for design optimization of electrical impedance spectroscopy systems

Abstract: The design of electrical impedance spectroscopy systems is driven by target applications. Moreover, system functions are inter-dependent. Thus, optimizing the design of each block without input from the others is impossible. Furthermore, simulating the whole system at transistor level is time consuming. This chapter presents a mixed-level simulation tool for the efficient development of bioimpedance spectroscopy measurement systems. We have developed *Verilog-AMS* models combined with transistor-level descriptions of several analog blocks. This allows us to run highly efficient system-level simulations for a whole circuit. We apply this tool to the design of an integrated circuit suitable for *in vivo* monitoring of bluefin tuna physiological parameters. In fact, the variation of biological processes offers significant insights into a host of key parameters such as growth, survival and reproduction. The obtained results show that the approach described in this chapter can reduce simulation time by 80 % while consistently reaching high simulation performances.

Keywords: bioimpedance spectroscopy, IC design, bluefin tuna, physiological parameter monitoring

1 Introduction

Electrical Impedance Spectroscopy (EIS) can be used to characterize matters governed by electrochemical processes. From material science to biosensing [1–3], the EIS range of applications is wide. In some of these applications, the system design is ruled by strong constraints. The simple choice between various potential system architectures is challenging. Moreover, system functions are inter-dependent. As a result, optimiz-

Acknowledgement: This study has been carried out in a project co-funded by IFREMER and Labex Numev.

Achraf Lamlih, Vincent Kerzérho, Serge Bernard, Fabien Soulier, Mariane Comte, Michel Renovell, LIRMM – University of Montpellier, 161 rue Ada, 34095 Monptellier Cedex 5, France, e-mail: Achraf.Lamlih@lirmm.fr
Tristan Rouyer, IFREMER, Avenue Jean Monnet, 34203 Sète, France
Sylvain Bonhommeau, IFREMER, Rue Jean Bertho, 97822 Le Port, France

https://doi.org/10.1515/9783110558920-008

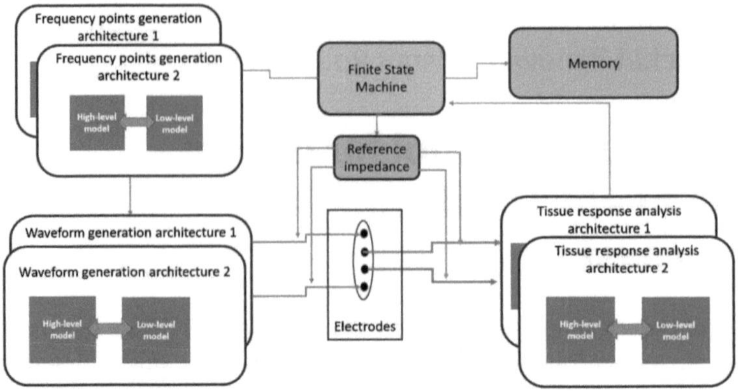

Fig. 1: The main blocks of the integrated circuit architecture.

ing the design of each block without input from the others is impossible. To address this issue, an efficient design strategy is needed.

The overall architecture consists of the main blocks presented in Fig. 1. The three principal system functions are the frequency components generation, the stimulation waveform construction and the system-under-test response analysis. The hardware implementation of each of these blocks differs depending on the application.

In general, three approaches are used for the design of integrated systems, i. e., the bottom-up, the top-down and the mixed-level approaches. The bottom-up approach consists in implementing each of the system blocks at transistor level, proceeding to the block verification, and then assembling all the blocks to build the system. This approach helps to achieve high performances at block level but without any system-level consideration. Thus, it is impossible to ensure optimal system-level performances. Moreover, a global simulation at transistor level is generally impractical because it is time consuming. Conversely, the top-down approach addresses architectural level first by describing each block by its functional model. This approach can optimize system-level specifications but without considering each block constraint. The high-level block requirements retrieved from the system-level simulation might be unsuitable at low level. The mixed-level approach combines high-level and low-level models for each block. The idea is to co-simulate some of the blocks at the transistor level with the others at high level. With this approach, we can address the inter-dependability of the system blocks to propose various optimized architectures, while customizing each block for better performances. Furthermore, by combining fast-simulated high-level models with accurate low-level models, we can drastically reduce the global simulation time. In this context, we develop a mixed-level simulation tool dedicated to help designers find the most suitable architecture for certain given system constraints.

Since the tool is intended to help design integrated circuit systems, the Cadence platform was chosen. We have developed a library of high-level and low-level models of the mainly used blocks for impedance spectroscopy from the literature [4–6]. For high-level models we use *Verilog-AMS*, which is a derivative of the *Verilog* hardware description language that enables designers to create analog and mixed-signal modules containing high-level behavioral descriptions of the system blocks. For the low-level model description, we use transistor-level models simulated by the *Cadence Virtuoso* suite.

We apply this tool to the design of an integrated circuit suitable for *in vivo* monitoring of bluefin tuna physiological parameters.

2 Bluefin tuna physiological parameter monitoring

Although marine ecosystem services are critical for human well-being, the necessary scientific information to conserve and manage them efficiently is currently lacking. A way to improve our understanding of marine species is to use implantable sensors for monitoring physiological parameters related to key biological processes such as feeding and spawning. Indeed, such processes offer significant insights into a host of key population parameters such as growth, survival and reproduction. For instance, a key physiological parameter is muscle fat content. The variation of this parameter in space and time is indeed related to the previously mentioned behaviors as fish feed to accumulate reserves, as the fat is used during reproduction.

2.1 System specifications related to the targeted application

The choice of the different elements constituting the architecture of the integrated circuit is related to the application. In our case, long monitoring periods in the wild are considered, typically 12 months. Therefore, the architecture should be characterized by low power consumption. As the device has to be implantable, solutions offering a low silicon surface area have to be preferred for the fish well-being.

Previous measurements conducted on bluefin tuna showed that the expected impedance values are of the order of 100 Ω at 50 KHz [7]. At higher frequencies, this value is expected to decrease. Using a handheld measurement device based on the AD5933 chip and a four-electrode configuration, bioimpedance spectroscopy ranging from 300 Hz to 100 KHz was performed on the bluefin tuna's back at the second dorsal fin level. The obtained results (Fig. 2) confirm the expected impedance range: from sub-Ohm levels at higher frequencies to hundreds of Ohms at lower frequencies.

The non-linear behavior of tissue could be linked to biological processes [8], thus allowing the collection of new data with substantial scientific potential. Therefore, the designed system must be able to measure non-linearity of the tissue.

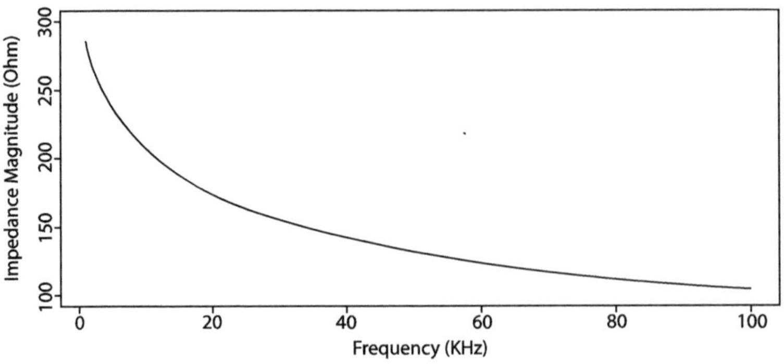

Fig. 2: Bluefin tuna impedance magnitude.

For a rigorous characterization of the biological tissue composition, the complex impedance needs to be analyzed over a large range of frequencies. Each type of tissue molecule has a different response to given frequencies. Three main frequency regions describing physiological processes have been identified [9]: the α dispersion (Hz–kHz), the β dispersion (kHz–MHz) and the γ dispersion (MHz–GHz). Since α and β dispersions are associated with the most relevant biological aspects [4] for our application and for integrated circuit (IC) design considerations, the exploration was limited to these two frequency regions: 500 Hz–10 MHz.

2.2 IC architecture definition

For the general principles of the measurement, current stimulation and voltage measurement were chosen, mainly for safety reasons since in that case the current passing through the biological tissue can be controlled. Also, a four-electrode configuration is used to get rid of the impedance at the interfaces.

To generate the α and β dispersions we use a phase-locked loop as a frequency multiplier. The low-frequency signal coming from a crystal oscillator is multiplied to get the higher frequency. On the negative feedback path, successive divisions are then performed to get the frequency points.

To analyze non-linearity of the tissue, sine waves have been chosen for the stimulus. In order to measure harmonics of the system (tissue), the generated sine wave should have the lowest total harmonic distortion THD possible, so that harmonic intermodulation does not affect the measured result. To cancel harmonics of the sine wave, scaled and delayed versions of the signal at the output of the frequency sweep generation block are added (Fig. 3). The scale coefficients and the phase shifts are computed so that the constructed signal matches the expression of a pure sine wave.

The result is then fed to a voltage-controlled current source (VCCS) that generates the stimulation current. The VCCS is designed in such a way to have high stability current in the targeted frequency range. Synchronous detection has been chosen as the

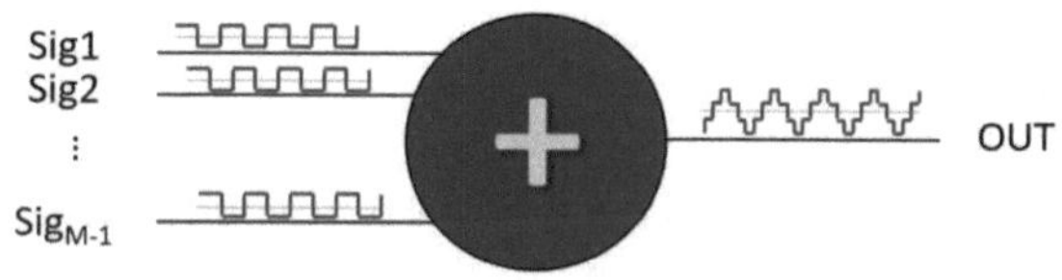

Fig. 3: Sine wave generation technique.

response analysis technique for its simplicity and its ability to single out the wanted frequency response even for the smallest AC signals obscured with noise. Since the frequency components of interest are translated into the baseband by the synchronous detection, the constraints on the analog-to-digital converter (ADC) are then relieved, and simple low-rate ADCs could be used to feed the memory with bit words representing the real and the imaginary parts of the impedance. A finite-state machine is used to control the different signals of the architecture, ensuring the sequentiality of the frequency sweep. Since the bluefin tuna is expected to migrate in several regions, the environmental parameters of the measurement (temperature, pressure, etc.) may vary drastically. A reference impedance is added to the overall architecture (Fig. 4) to compensate for the system drifts.

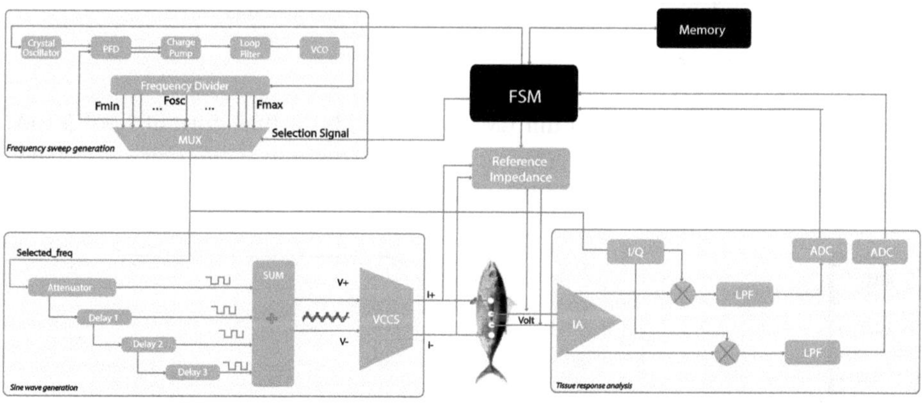

Fig. 4: The system-on-chip architecture.

2.3 Mixed-level simulation of the architecture

The VCCS is a critical component in bioimpedance spectroscopy architectures. It should deliver a high-accuracy current to the tissue under test over the frequency range of use. Unfortunately, at higher frequencies, the stray capacitance shunts the output impedance of the current driver, which causes its performance to decrease. To illustrate the mixed-level simulation approach on the proposed architecture, we chose to implement the VCCS at the transistor level to target its performances and the other blocks at high level using Verilog-AMS.

2.3.1 Voltage controlled current source design

The current driver is implemented using a symmetrical operational transconductance amplifier (OTA) (Fig. 5). It is composed of a differential pair (M1, M2) driving two loads (M3, M4) connected as diodes, three current mirrors (M3–M5, M4–M6, M7–M8) and a current source M9 [10]. The first stage (M10, M11, M12, M13) is designed to bias the OTA with a current I_b.

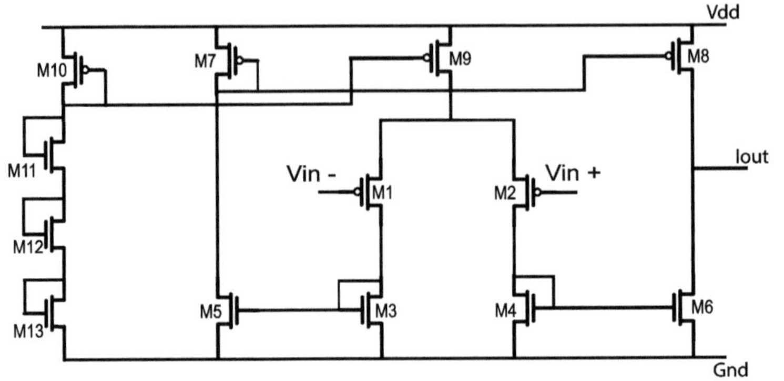

Fig. 5: The current driver architecture.

The OTA is designed with a 180-nm CMOS process, for a bias current I_b of 50 μA, a dynamic input range of ±0.3 V (Fig. 6) around a common mode voltage of 0.6 V, a transconductance of 200 μS and a bandwidth of 10 MHz.

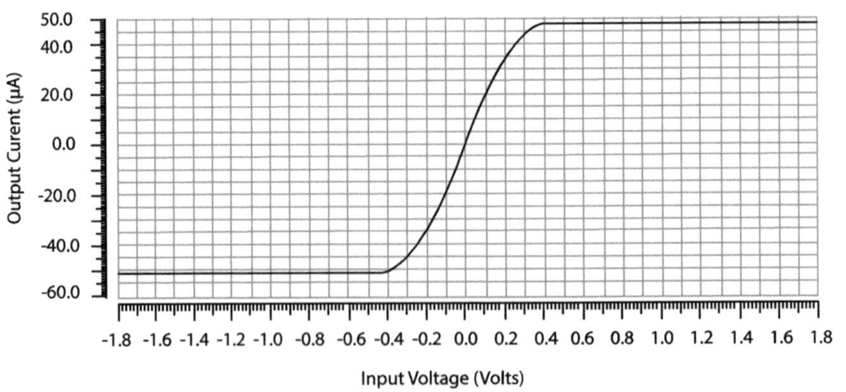

Fig. 6: The OTA differential DC input–output transfer curve.

The achieved output impedance of the current driver is 270 kΩ at frequencies below 5 MHz; it drops to 259 kΩ at 10 MHz (Fig. 7). The output impedance value could be increased by cascoding the output stage, for example, at the price of decreasing the output swing of the current driver. Although these values of output impedance are

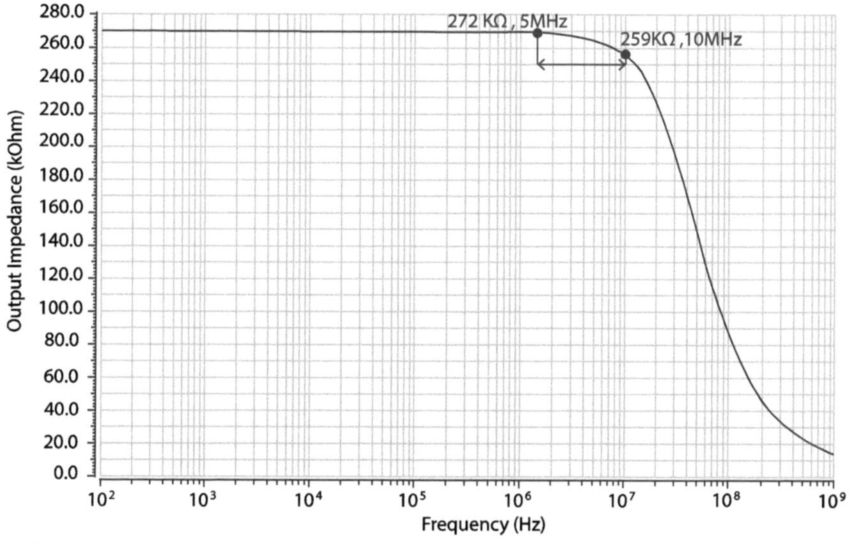

Fig. 7: The current driver output impedance vs frequency, with a load of 1 kΩ/3 pF.

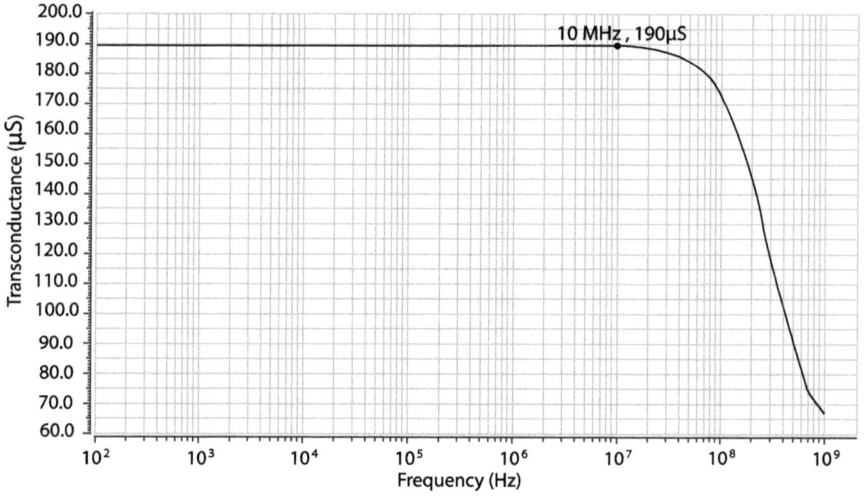

Fig. 8: The current driver transconductance vs frequency, with a load of 1 kΩ/3 pF.

not high, they are suitable for our application since the expected impedance of the bluefin tuna does not exceed 100 Ohms (Fig. 2). The achieved transconductance value is 190 μS and remains stable over the desired bandwidth (Fig. 8).

Monte Carlo simulation was performed on the transconductance of the OTA for 1000 runs with both process and mismatch; the results are shown in Fig. 9. The transconductance has a mean value of 190.64 μS and a standard deviation of 3.68 μS. The value of the standard deviation could be decreased by increasing the dimensions

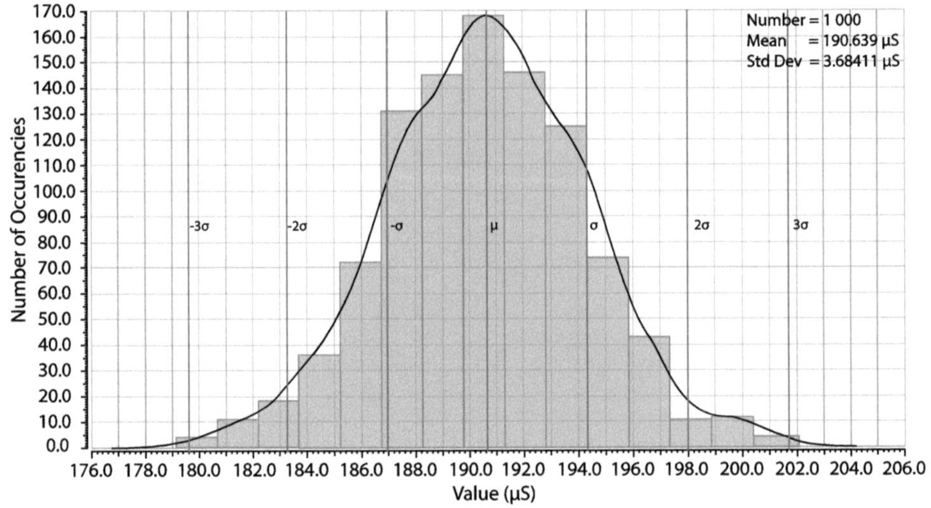

Fig. 9: Monte Carlo simulation of the transconductance of the current source.

of the current mirrors of the output stage, at the price of decreasing the bandwidth due to larger parasitic capacitances. Since 97 % of the runs are within 3.8 % of the deviation from the mean value, the design is considered as robust under process variation and device mismatch, and the transistor dimensions are kept unchanged.

2.3.2 Simulation performances

To illustrate the efficiency of the mixed-level approach, two transient simulations of the overall architecture described below were launched. The first simulation contained the current source designed at the transistor level and the other blocks at high level. The second simulation contained all the blocks of the architecture at high level; a realistic Verilog-AMS model of the current source was used.

The Verilog-AMS model takes into account the values of the transconductance and the output resistance extracted from the simulation results presented in the previous part. The model could also be used to evaluate at high level the impact of process variation and device mismatch by using the results of the Monte Carlo simulation. In fact, the impedance error introduced by process and mismatch variation is ±1.9 % within one standard deviation from the mean value.

Simulations were performed on a machine with 25 cores (2 GHz) and 263 Gb of memory. The entire high-level simulation with the Verilog-AMS model of the current source took 15 minutes. On the other hand, the simulation with the current source at transistor level and the other blocks at high level took 75 minutes. The obtained results show that the mixed-level approach described in this chapter can reduce the simulation time by 80 % while consistently reaching high performances for each block.

3 Conclusion

This chapter presents a mixed-level simulation tool to support the design of bio-impedance spectroscopy ICs. The mixed-level simulation addresses the inter-dependability of the different blocks of the architecture, while enabling the designers to test different architecture solutions. Furthermore, the tool permits to go further in the design customization, allowing designers to enhance system performances with the significant advantage of having reduced simulation times. Moreover, this approach could also be useful to estimate the influence of process variation and device mismatch on the overall architecture. The influence of the measurement environment could also be estimated. In fact, Verilog-AMS blocks modeling sources of inaccuracies (such as the effect of the temperature and flicker noise) could simply be added to the model, permitting the evaluation of the impact and if needed refining the architecture for robustness.

Bibliography

[1] P. Aberg, N. Ingrid, J. Hansson, P. Geladi, U. Holmgren, and S. Ollmar, "Skin cancer identification using multifrequency electrical impedance", IEEE Transactions on Bio-Medical Engineering, 2004.

[2] T. Dai and A. Adler, "In vivo blood characterization from bioimpedance spectroscopy of blood pooling", IEEE Transactions on Instrumentation and Measurement, 2009.

[3] S. Grimnes and O. G. Martinsen, "Bioimpedance and Bioelectricity Basics".

[4] H. M. Dastjerdi, R. Soltanzadeh, and H. Rabbani, "Designing and implementing bioimpedance spectroscopy device by measuring impedance in a mouse tissue", Journal of Medical Signals and Sensors, 2013.

[5] Y. Long, B. Joonsung, L. Seulki, and Y. Hoi-Jun, "A 3.9 mW 25-electrode reconfigured sensor for wearable cardiac monitoring system", IEEE Journal of Solid-State Circuits, 2011.

[6] S. Rodriguez, S. Ollmar, M. Waqar, and A. Rusu, "A batteryless sensor ASIC for implantable bio-impedance applications", IEEE Transactions on Biomedical Circuits and Systems, 2016.

[7] J. Willis and A. J. Hobday, "Application of bioelectrical impedance analysis as a method for estimating composition and metabolic condition of southern bluefin tuna (Thunnus maccoyii) during conventional tagging", 2008.

[8] T. Yamamoto and Y. Yamamoto, "Non-linear electrical properties of skin in the low frequency range, Medical and Biological Engineering and Computing", 1981.

[9] H. P. Schwan, "Electrical properties of tissue and cell suspensions".

[10] M. C. Sansen, "Analog Design Essentials", The Springer International Series in Engineering and Computer Science, 2016.

René Schmidt, Stephan Blokzyl, and Wolfram Hardt

A highly scalable FPGA implementation for cross-correlation with up-sampling support

Abstract: Cross-correlation is a fundamental algorithm which is used in many fields of digital signal processing. One important application is sound source localization. For this purpose, actual research shows a significant improvement of localization accuracy when using up-sampling-based algorithms. These state-of-the-art approaches are challenging for embedded realizations as up-sampling in combination with high sample rates and measurement window lengths leads to huge amounts of data to be processed and stored.

This chapter presents hardware acceleration techniques for cross-correlation with up-sampling support. The work introduces a novel processing architecture which addresses large-scale memory consumption compared to conventional solutions. The novel design introduces a highly scalable field programmable gate array realization with linear memory complexity and an interpolation factor-independent runtime.

Keywords: mathematics algorithms, digital signal processing, splines, field programmable gate arrays

1 Introduction

Cross-correlation algorithms are widely used in various fields of digital signal processing. Typical applications comprise image processing [1], audio signal processing [2], impedance spectroscopy [3] and sound source localization [2]. Especially for sound source localization, approaches like time difference of arrival (TDOA) are used, a technique to calculate the position of a sound source by computing the delay between incoming sound signals of spatially divided microphones [4]. The cross-correlation is used to calculate the audio signal lag (signal time difference). There are two common models for cross-correlation. The general definition is

$$u(t) \otimes v(t) = \int u^*(\tau)\, v(\tau + t) d\tau. \tag{1}$$

Since digital signals are discrete and real-valued, the definition (1) can be written

$$u(t) \otimes v(t) = \sum u(\tau) v(\tau + t). \tag{2}$$

René Schmidt, Stephan Blokzyl, Wolfram Hardt, Chemnitz University of Technology, Faculty of Computer Science, Professorship of Computer Engineering, Chemnitz, Germany, e-mail: rene.schmidt@informatik.tu-chemnitz.de

https://doi.org/10.1515/9783110558920-009

Alternatively, the cross-correlation can be calculated by using convolution, given by

$$u(t) * v(t) = \int u(\tau)v(\tau + t)d\tau. \tag{3}$$

An equivalent description is

$$u(t) * v(t) = \mathcal{F}^{-1}(\mathcal{F}(u(t)) * \mathcal{F}(v(t))). \tag{4}$$

The calculation (4) has recently been used in micro-controller-based realizations, since the fast Fourier transform (FFT) $\mathcal{F}$ can be implemented efficiently, in contrast to the traditional estimation (2), which has a runtime of $O(n\tau_{\max}) = O(n^2)$ [5]. FFT calculation is also appropriate for field programmable gate array (FPGA)-based solutions [6]. Hence, predesigned FFT IP cores are resource-intensive and restricted in the number of samples (compare Section 3).

Two major properties specify TDOA localization accuracy. One is the base distance between the microphones, which is proportional to the measurement precision. Small base distances cause high measurement uncertainty [7]. However, the base distance cannot always be adapted to a length which enables the required measurement accuracy. The second feature is the sampling rate. The measurement error reduces with a higher sampling rate, but sample rate incrementation leads to a higher data volume to be processed. With a measurement interval of 20 ms and a default microphone sample rate of 44.1 kHz, 882 values have to be correlated each measurement cycle. This amount of data is not a problem, but with increasing sample rate the computational effort raises significantly, which is challenging for embedded devices.

The sample rate can be increased by using faster ADCs or by generating additional artificial data with up-sampling. Kan has presented an up-sampling method to improve localization accuracy [2]. The challenge can be illustrated with the following example. An up-sampling rate of 64 together with a sample rate of 44.1 kHz leads to an increased sample rate of approximately 2.8 MHz and 56,448 data samples to be computed. The restriction of, e. g., the Xilinx reference FFT IP core to 65,536 data samples [8] allows a maximum sample rate of approximately 3.3 MHz. Higher sample rates cannot be realized without measurement window reduction.

This chapter is structured as follows. Section 2 introduces related work and applications and validates state-of-the-art cross-correlation methodologies with respect to embedded realizations. Section 3 gives a review of an FPGA-based reference design that is based on common cross-correlation as presented in standard literature. Section 4 presents a novel architecture for a highly scalable FPGA implementation of up-sampling supporting cross-correlation. Finally, Sections 5 and 6 conclude the chapter with the results and benefits of the proposed solution.

2 Related work

There is a considerable amount of literature discussing versatile applications of cross-correlation. The proposed approaches apply software-based cross-correlation specified and optimized for different measurement applications. Venable presented a work in 2006 using cross-correlation for the estimation of peptide molecular weights from tandem mass spectra [9]. With the tool SEQUEST, a cross-correlation-based template matching approach is used for spectrogram comparison [10].

In the field of image exploitation, the fast normalized cross-correlation has been established as novel template matching (TM) method [1, 11]. The TM-based cross-correlation has been used for medical applications like motion recognition in blood flow imaging [12] and frontotemporal dementia [13] and for quality assurance in industrial production processes like defect detection of electronic devices [14]. Other applications use cross-correlation to calculate time differences between two related signals, like sound source localization methods [4], as mentioned above.

The listed algorithms are realized in software and cause disadvantageous high computational costs [5], which is challenging in embedded applications. Therefore, FPGA technology is applied to accelerate computation with the help of parallel processing strategies. FPGA-based cross-correlation architectures, as proposed in the literature, can be classified into two main categories: solutions using discrete cross-correlation and methods applying FFT. Approaches from the first category solve equation (2) and implement a parallel or serial architecture [15] as depicted in Fig. 1. These architectures are used in the field of orthogonal frequency division multiplexing [16] and other high-performance applications [17]. The serial structure provides a high processing speed, since each sample X needs to be a processed one. However, resource consumption rises linearly with the amount of samples per measurement window. As each measurement sample requires a single multiplier for cross-correlation, e. g., 20 samples consume 20 and 20,000 samples consume 20,000 multipliers. The relation between the measurement sample and multipliers is linear. With 220 DSP cores for multiplication, as provided by state-of-the-art FPGAs,[1] the proposed implementation is not feasible for measurement scenarios with high sample rates or large-scale measurement windows.

In contrast to the serial approach, the parallel architecture saves resources, since the calculation needs to be executed for each possible lag [15]. However, the parallel architecture consumes a big amount of multipliers to realize cross-correlation for high sample rates. As the count of multipliers is limited in FPGAs, both serial and parallel solutions are not feasible for high sample rates and large measurement windows. Thus, state-of-the-art FPGA-based solutions [6, 18] use FFT-based cross-correlation realization with convolution (see equation (4)). Fig. 2 shows the appropriate hardware

1 E. g., Xilinx Zynq®-7000 All Programmable SoC XC7Z020-CLG484.

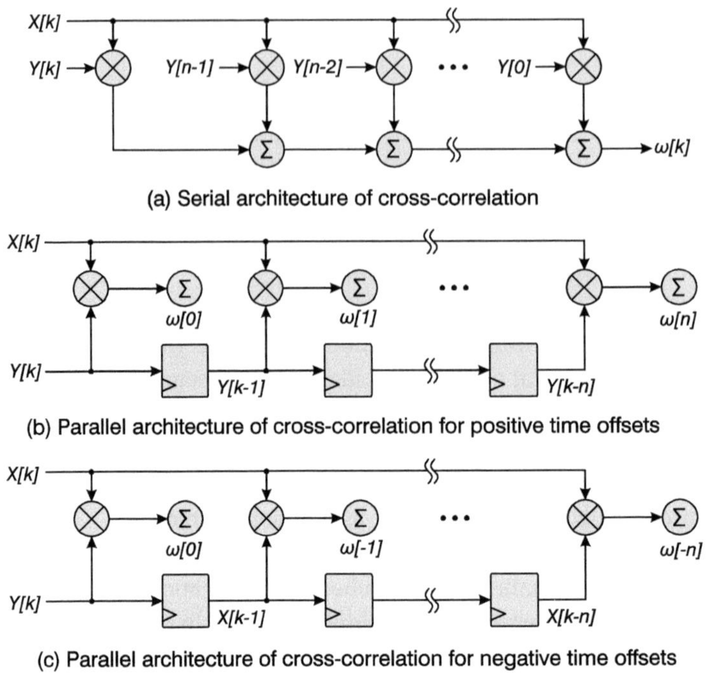

Fig. 1: Comparison of common FPGA architectures for direct calculation of discrete cross-correlation. (a) The stepwise, serial realization of cross-correlation. All X-samples have to be processed for the final cross-correlation result. (b) and (c) The highly parallel architecture for cross-correlation which provides a result update every clock cycle.

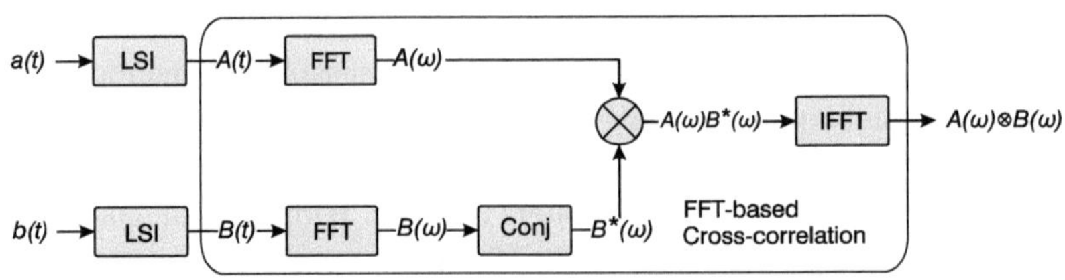

Fig. 2: FFT-based cross-correlation architecture with linear spline interpolation-based preprocessing.

implementation applying parallel FFT calculation. Various IP cores for efficient, high-performance FFT computation are available, easy to use, and optimized for specific FPGA devices, which makes them comfortable for modern FPGA designs.

A literature survey and manifold application scenarios prove the cross-correlation is a potential method for a wide range of commercial, industrial and research fields. The high computational effort and significant computational resource utilization represent the major drawbacks of state-of-the-art methodologies, especially for embedded realizations. The following analysis discusses standard FPGA-based cross-correlation in detail and derives an approach facing specific resource limitations of FPGA technology while providing higher computational performance compared to conventional cross-correlation approaches.

3 Reference design

The FPGA-based hardware-accelerated design for cross-correlation, depicted in Fig. 2, has been implemented on an evaluation board with Xilinx Zynq®-7000 SoC to compare and assess different realization alternatives and implementation techniques. Two parallel FFT modules simultaneously transform two time domain input signals $A(t)$ and $B(t)$ from spatially separated microphones into a frequency domain. The design substitutes time domain cross-correlation with multiplication in the frequency domain of signal A and the complex conjugated of signal B.

To support the up-sampling methodology by Kan, which improves the measurement accuracy (compare Section 1), both signal $a(t)$ and signal $b(t)$ have to be interpolated to $A(t)$ and $B(t)$ prior to FFT. A common interpolation method is, e. g., linear spline interpolation (LSI). The inverse fast Fourier transform (IFFT) converts the multiplication result back to the time domain, which represents the cross-correlation result. The most prominent peak of the cross-correlation result characterizes the time offset between signal A and signal B.

The two FFTs have been implemented as dual-channel FFT, which provides the best resource optimization provided by the vendor Xilinx. Fig. 3 (a) shows the appropriate utilization of Block RAM (BRAM) and DSP resources in relation to the measurement window length. The number of DSP slices is constant and independent of the measurement window size. The applied Radix-2 FFT algorithm [19] has a fixed structure and does not change with measurement window variation. The BRAM count increases linearly and is proportional to the measurement window length. The BRAM components are used as data buffer only and not applied for calculation. The memory increases by expanding the measurement window length and vice versa.

The Radix-2 algorithm uses a fixed amount of flip-flops (FFs) and lookup tables (LUTs) for FFT. Hence, measurement window length variation results in a logarithmic relation of logic resource consumption (see Fig. 3 (b)). A larger measurement window size requires a larger address space and additional glue logic, but the growth rate for buffer control logic consumption reduces with the increasing buffer size.

As a result, the processing time increases linearly with the measurement window size (see Fig. 4). With a logic utilization less than 10 % on a Zedboard with Zynq®-7000 All Programmable SoC XC7Z020-CLG484, the method is very resource saving. The amount of FFs and LUTs is logarithmically related to the number of samples per measurement window to be processed. However, the amount of used memory (BRAM) is very high and the reference cross-correlation architecture cannot be realized on small,

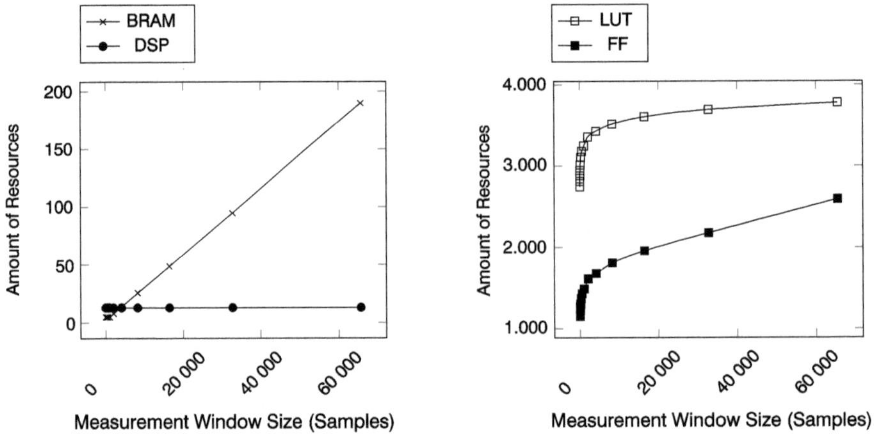

(a) Linear dependency of BRAM utilization as a function of the sample count in the measurement window. The amount of applied DSP cores is independent of the amount of samples in the measurement window.

(b) Logarithmic dependency of both FF and LUT utilization as a function of the amount of samples per measurement window.

Fig. 3: Logic resources utilization for the reference design depicted in Fig. 2 on a Zedboard with Xilinx Zynq®-7000 All Programmable SoC XC7Z020-CLG484.

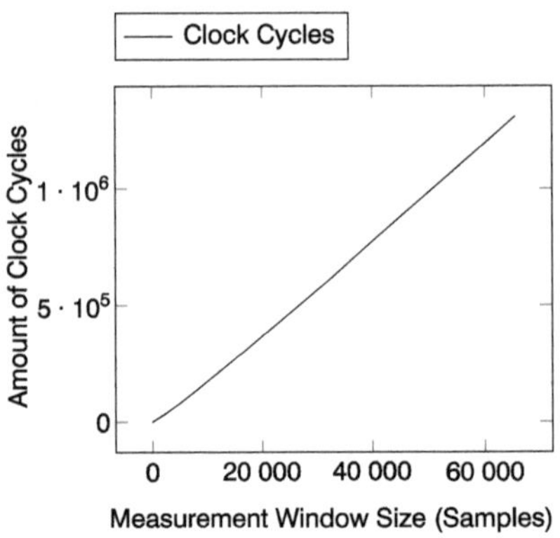

Fig. 4: Linearly rising FPGA processing time for the reference cross-correlation realization depicted in Fig. 2 as a function of the measurement window size.

low-budges FPGA devices with very limited BRAM resources. As the cross-correlation is just a single part of the overall solution, it shall not occupy the majority of BRAM resources of the host device.

Especially the storage of additional data caused by linear spline interpolation consumes a lot of BRAM resources. As the interpolation data are artificial and do not represent real measurements, a strategy to prevent the buffering of interpolation data will improve the BRAM utilization. While the reference FFT IP core buffers internally all data to process, the availability of interpolation data is essential for the convolution approach. Furthermore, the FFT core uses a measurement window size with a power of two step width only. If fewer data are processed the remaining window share is filled with zeros, which causes unnecessary computational effort. The maximum lag can be determined for most of the TDOA use cases. Hence, it is not necessary to calculate the complete cross-correlation, which saves calculation time. The hereinafter highly scalable FPGA implementation addresses these disadvantages and faces the trade-off between resource limitation and measurement accuracy. The method uses fixed-point arithmetic and provides high up-sampling capabilities with linear spline calculation.

4 Novel acceleration architecture

The novel acceleration architecture for FPGA-based cross-correlation consists of 14 different components (see Fig. 5). The first two parts are two True Dual Port Block-RAMs (BRAM X, BRAM Y) which buffer sampled input data that have to be processed. The first port is assigned for filling the BRAM and the second port provides access to the connected processing modules. The subsequent component is the data access (DA) block. This module controls input data readout from the BRAMs and handles interfaces and address mapping.

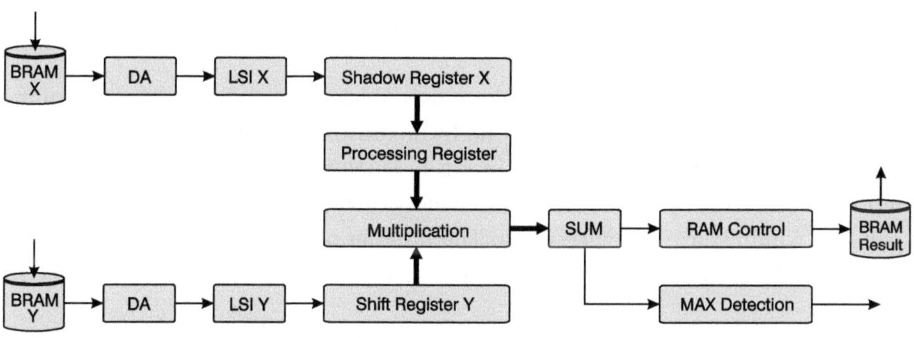

Fig. 5: Novel acceleration architecture for FPGA-based cross-correlation.

Next, the LSI block stores two subsequent values and calculates interpolation values depending on the interpolation factor ξ. The LSI block for x-values feeds a shadow register, which is organized as a shift register with a length of ξ. The interpolation block of the y-values feeds a second shift register with equal length ξ. The initialization value of LSI Y is for each interpolation step $X_n - \tau_{max}$ and generates $k = 2\tau_{max}+1$ values, while the next ξ x-values are calculated simultaneously; τ_{max} describes the maximum delay (in clock cycles) of the cross-correlation given by $\tau_{max} = f(F_s, \xi)$ with sample rate F_s. After each processing step of all k values, the shadow register X is copied to the processing register. The ξ interpolation values in shift register Y are multiplied in parallel with all ξ interpolation values in the processing register, which results in the products $\omega_i(\tau)$ with $\{i \in \mathbb{N} \mid 0 \leq i < \xi\}$. All multiplication results $\omega_i(\tau)$ are finally summed up to $\Omega(\tau)$ in the SUM module.

In reference to equation (2), the interim result from the previous processing step for X_{n-1} is prepared by the BRAM control block, transferred to the SUM block, and accumulated in the processing step of X_n. The BRAM control block stores the updated X_n in BRAM result and handles the storage access. The maximum detection index and the maximum detection value are stored by the MAX detection block after each summation step. The procedure is repeated for all $(N \cdot \xi)$ x-values, where N is the amount of data per window. Since the multiplication generates big numbers and the range limit of integers is reached soon, the whole processing is done with fixed-point arithmetic. The width of the input data is defined as 32 bits, with 16 bits for the integer part and 16 bits for the fractional part.

5 Results

The algorithm calculates for each x-value ξ interpolation values and processes them in parallel. The computation is done $2\tau_{\max} + 1$ times, which leads to a runtime of $O(2N\tau_{\max} + 1)$, depending on the according sample rate. The runtime O is independent of the interpolation factor ξ since all interpolation values are calculated at the same point of time.

Logic resource consumption of the proposed design depends on the value of interpolation factor ξ. FF, LUT, DSP slice and BRAM utilization is proportional to ξ with linear hardware complexity as shown in Fig. 6. Furthermore, both FF and LUT consumption are affected by the measurement window size and grow logarithmically with expanding window size and vice versa. The number of required DSP slices is independent of the measurement window length as all interpolation steps are executed in parallel.

The BRAM utilization is linear and strongly linked to the runtime and changes with sample rate, measurement windows size and interpolation factor modifications (see Fig. 7). Window size and sample rate changes affect the two data input buffers (BRAM X/Y) and the BRAM for results, while a modification of ξ affects the result BRAM only. The growth of BRAM consumption is moderate as existing memory utilization rises up to 100 % before a new memory block will be activated. The proposed architecture uses significantly less BRAM resources compared to the FFT IP core-based reference design. E. g., a hardware realization with an interpolation factor of $\xi = 64$ requires only ˜60 % more FFs, ˜10 % more LUTs, and five times more DSP slices, but saves more than 97 % of BRAM resources for 65k samples (maximum of reference design). With 18 FFs (0.3 %) and 60 LUTs (2 %) more, the proposed design is able to process 262k samples (equal number of DSP slices and four times the BRAM consumption), which equals four times the performance compared to the reference design. The

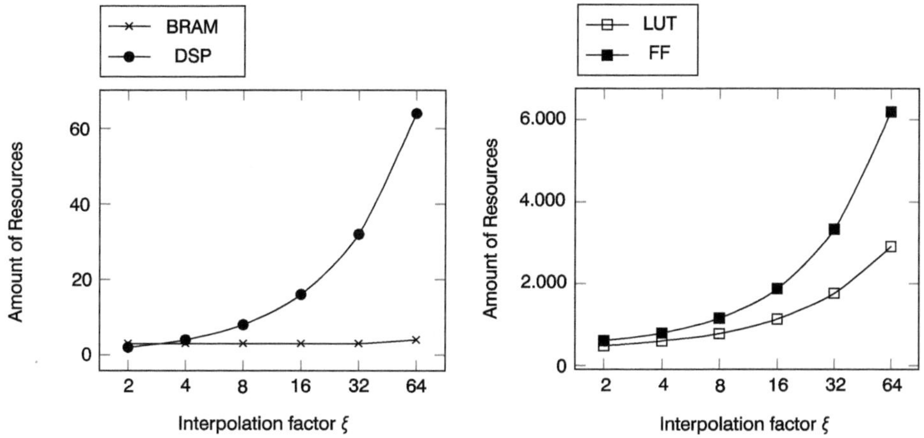

(a) Linear relationship between interpolation factor ξ and DSP slice consumption. BRAM utilization is independent of ξ.

(b) Linear relationship between interpolation factor ξ and required FFs and LUTs.

Fig. 6: Influence of the interpolation factor ξ on resource utilization of the proposed acceleration architecture for FPGA-based cross-correlation.

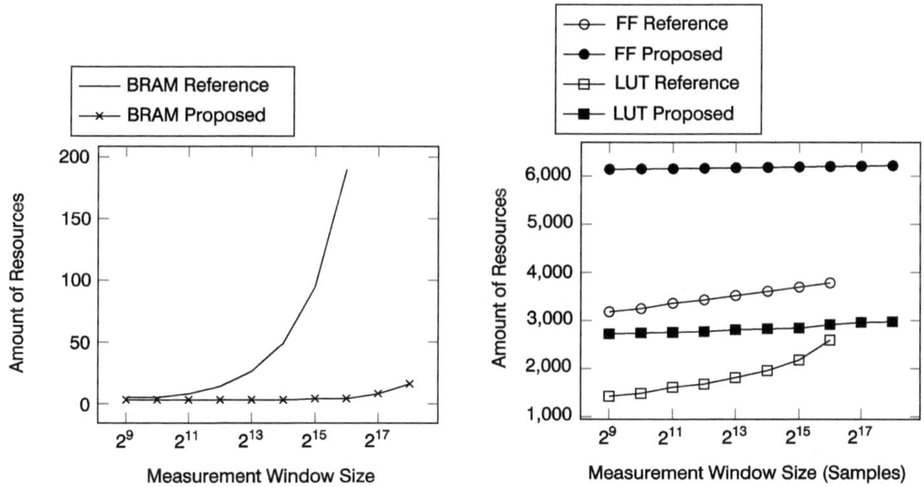

(a) Comparison of BRAM utilization between the reference design and the proposed design.

(b) Comparison of FF and LUT utilization between the reference design and the proposed design.

Fig. 7: Comparison of resource consumption of the reference and proposed designs for interpolation factor $\xi = 64$.

proposed design does not store intermediate multiplication results, which leads to a significant reduction of BRAM resource consumption compared to the reference design. Furthermore, the measurement window length and sample rate do not affect the overall processing part of the design. FF and LUT consumption is nearly constant and rises minimally for glue logic only, which is negligible compared to the available FF and LUT resources on modern FPGA devices. The independence of logic resource consumption from different measurement configurations leads to a beneficial scalability of the overall architecture. Additionally, the novel method provides a measurement window configuration of arbitrary size, which is only limited by the available logic resources of the applied FPGA.

6 Conclusion and future work

The proposed FPGA design provides a highly scalable FPGA architecture for hardware-accelerated cross-correlation with up-sampling support. The novel approach has a runtime of $O(2N\tau_{max})$ and a memory complexity of $O(2N+3\xi+2\tau_{max})$. The up-sampling is realized by LSI and characterized by an interpolation factor ξ. The modification of the interpolation depth affects the consumption of DSP slices and FFs only and does not influence the overall processing time, which is solely coupled to window size and sample rate.

The novel approach enables arbitrary and unlimited measurement window lengths, which allows an extension of the maximum measurement window size compared to the reference design. The proposed solution requires significantly less BRAM and LUT resources for both high sample rates and large-scale measurement window configurations. The presented architecture occupies slightly more FFs, which is negligible with respect to the available FFs on modern FPGA devices.

Future activities focus on the analysis of time restrictions to ensure real-time capabilities of the design. Furthermore, techniques for a significant reduction of DSP slices have to be developed.

Bibliography

[1] J. Lewis, “Fast normalized cross-correlation”, Vision Interface, vol. 10, pp. 120–123, 1995.

[2] Y. Kan, P. Wang, F. Zha, et al., “Passive acoustic source localization at a low sampling rate based on a five-element cross microphone array,” Sensors, vol. 15, no. 6, pp. 13326–13347, 2015.

[3] S. Gawad, T. Sun, N. G. Green, et al., “Impedance spectroscopy using maximum length sequences: application to single cell analysis”, Review of Scientific Instruments, vol. 78, no. 5, 054301, 2007.

[4] C. H. Knapp and G. C. Carter, "The generalized correlation method for estimation of time delay", IEEE Transactions on Acoustics, Speech, and Signal Processing, vol. 24, no. 4, pp. 320–327, 1976.

[5] D. Lyon, "The discrete fourier transform, part 6: cross-correlation.", Journal of Object Technology, vol. 9, no. 2, pp. 17–22, 2010.

[6] D. Nguyen, P. Aarabi, and A. Sheikholeslami, "Real-time sound localization using field-programmable gate arrays", In: Acoustics, Speech, and Signal Processing, 2003. Proceedings. (ICASSP'03). 2003 IEEE International Conference on, IEEE, vol. 2, 2003, pp. II–573.

[7] R. Schmidt and W. Hardt, "Schallquellenlokalisierung zur sprecheridentifizierung fur multi-user dialogschnittstellen", In: Studierendensymposium Informatik 2016 der TU Chemnitz, TU Chemnitz, vol. 1, 2016, pp. 105110.

[8] L. I. P. Guide, "Fast fourier transform v9. 0".

[9] J. D. Venable, T. Xu, D. Cociorva, et al., "Cross-correlation algorithm for calculation of peptide molecular weight from tandem mass spectra", Analytical Chemistry, vol. 78, no. 6, pp. 1921–1929, 2006.

[10] J. K. Eng, B. Fischer, J. Grossmann, et al., "A fast sequest cross correlation algorithm", Journal of Proteome Research, vol. 7, no. 10, pp. 4598–4602, 2008.

[11] J.-C. Yoo and T. H. Han, "Fast normalized cross-correlation", Circuits, Systems, and Signal Processing, vol. 28, no. 6, pp. 819–843, 2009.

[12] J. Luo and E. E. Konofagou, "A fast normalized cross-correlation calculation method for motion estimation", IEEE Transactions on Ultrasonics, Ferroelectrics, and Frequency Control, vol. 57, no. 6, pp. 1347–1357, 2010.

[13] B. B. Avants, C. L. Epstein, M. Grossman, et al., "Symmetric diffeomorphic image registration with cross-correlation: evaluating automated labeling of elderly and neurodegenerative brain", Medical Image Analysis, vol. 12, no. 1, pp. 26–41, 2008.

[14] D.-M. Tsai and C.-T. Lin, "Fast normalized cross correlation for defect detection", Pattern Recognition Letters, vol. 24, no. 15, pp. 2625–2631, 2003.

[15] A. Fort, J.-W. Weijers, V. Derudder, et al., "A performance and complexity comparison of auto-correlation and cross-correlation for ofdm burst synchronization", In: Acoustics, Speech, and Signal Processing, 2003. Proceedings. (ICASSP'03). 2003 IEEE International Conference on, IEEE, vol. 2, 2003, pp. II–341.

[16] M. Chen, J. He, and L. Chen, "Real-time optical ofdm long-reach pon system over 100 km ssmf using a directly modulated dfb laser", Journal of Optical Communications and Networking, vol. 6, no. 1, pp. 18–25, 2014.

[17] B. Von Herzen, "Signal processing at 250 mHz using high-performance fpga's", IEEE Transactions on Very Large Scale Integration (VLSI) Systems, vol. 6, no. 2, pp. 238–246, 1998.

[18] J. Pang and J. Starzyk, "Fast direct gps signal acquisition using fpga", In: ECCTD, Krakow:[sn], vol. 287, 2003.

[19] S. Goedecker, "Fast radix 2, 3, 4, and 5 kernels for fast fourier transformations on computers with overlapping multiply-add instructions", SIAM Journal on Scientific Computing, vol. 18, no. 6, pp. 1605–1611, 1997.

Ahmed Fendri, Racem Jribi, Olfa Kanoun, and Hamadi Ghariani

Interface circuit for oil quality assessment considering dielectric losses and stray capacitances

Abstract: Dielectric spectroscopy is becoming a popular technique in addition to the chemical and optical methods for oil quality assessment. It is based on the measurement of the variation of the dielectric properties of oil corresponding to the change of its chemical and physical properties. In this investigation an interface circuit for the measurement of the dielectric properties of oil is proposed using the capacitance-to-voltage conversion technique. The circuit measures the variation of capacitance of the sensor. The output voltage of the circuit is a DC voltage proportional to the capacitance of the sensor with a resolution of 5 mV/pF. The differential technique was used to eliminate stray capacitances caused by the circuit. Regression analysis has been carried out to describe a mathematical relation between the change of the capacitance and the output voltage. Simulations show a maximum relative deviation of 0.03 % for capacitance in the range of 100 pF to 1000 pF while experimental results show a maximum relative deviation of 2 % for the same capacitance range. The effects of stray capacitances and dielectric losses are also investigated. Stray capacitances decrease the measurement accuracy to 0.1 %, while the parallel resistance, which represents the dielectric losses does not affect the accuracy, if its value is higher than 10 MΩ. For resistance values lower than 10 MΩ the circuit is not stable and it is threfore not possible to estimate the capacitance value using the output voltage.

Keywords: dielectric spectroscopy, capacitive sensor, capacitance measurement, capacitance-to-voltage conversion

1 Introduction

The measurement of small capacitance changes is very important in industrial applications and in particularly for the measurement of dielectric properties of liquids such as oils. In fact, the measurement of a sensor capacitance dipped in oil can be used to

Ahmed Fendri, Olfa Kanoun, Chair for Measurements and Sensor Technology, Chemnitz University of Technology, Reichenhainer Strasse 70, 09126, Chemnitz, Germany, e-mail: ahmed.fendri@s2013.tu-chemnitz.de
Racem Jribi, National Engineering School of Gabes, Cabès, Tunisia
Hamadi Ghariani, National School of Engineers of Sfax, University of Sfax, Sfax, Tunisia

https://doi.org/10.1515/9783110558920-010

measure the dielectric properties of the oil. Many investigations used this technique to characterize edible oils [1], detect the adulteration [2] and estimate the water content [3] or the total polar components in fried oils [4]. A typical measurement setup consists of a capacitive sensor connected to an impedance analyzer or an LCR meter, which measures the capacitance of the sensor and thus deduces the dielectric properties. Such a measurement setup is accurate and simple to use but it has some limitation since it is not designed for online or mobile applications but only for laboratory measurement.

Since new investigation are not focusing only in laboratory conditions but also in online industrial applications where the ideal conditions are not guaranteed and a smaller mobile setup with low energy consumption is required. It consists of an embedded measurement circuit connected to a well designed capacitive sensor. The measurement circuit converts the capacitance to measurable data (voltage, frequency, time, etc.), while the sensor is used in such a way that the change in the dielectric properties will cause a change in the capacitance.

The capacitance measurement circuits can be classified into three main categories according to the type of conversion the circuit does:
- capacitance to time,
- capacitance to frequency,
- capacitance to voltage.

The capacitance-to-time conversion includes the charging/discharging and phase shift techniques. The working principle of the charging/discharging technique is simple; the capacitance to be measured is charged during a known period of time and then the required time to discharge the capacitor is used to measure its value. It is also possible to measure the capacitance during the charging process by measuring the required time to charge the capacitor to a certain voltage. CMOS switches are usually used as switches to charge and discharge the capacitor; in fact the use of switches increases the stray capacitances and hence the measurement errors [5, 6]. The second technique is the capacitance-to-phase angle conversion; one of the capacitor electrodes is connected to an AC voltage source while the second electrode is connected to a charge amplifier and a compensating capacitor excited by a compensating signal. The use of this compensation permits a gradual phase shift in response to the change of the capacitance. This enables the circuit to measure very small capacitance changes but in a very limited range. The phase angle of the output signal of the charge amplifier is compared to that of the input signal and the difference in phase is converted to time using a Schmitt trigger [7].

Another techniques to measure capacitance is to convert its value to a periodic signal. Oscillation and resonance circuits are the main two types of the capacitance-to-frequency conversion technique. The oscillator produces a periodic signal with a frequency proportional to the capacitance C_x [8]. The resonance technique is based on the use of a parallel or serial inductance and resistance to the capacitance to be measured in order to create a resonator. The circuit is excited by an AC voltage with a variable frequency; the voltage and the current in the circuit should be measured at each frequency until the resonance happens where the current or the voltage is maximum, in case of a serial or parallel circuit, respectively [9].

The capacitance-to-voltage type includes mainly two techniques of conversion: the transducer and the I/Q modulation. The transducer circuit is simple and fast; it consists basically of a current-to-voltage converter, a rectifier and a low-pass filter in such a way that it provides a DC output voltage with an amplitude depending on the value of the capacitance of the sensor [10, 11]. This type of circuit does not consider the effect of dielectric losses existing in capacitive sensors. The I/Q modulation circuit considers the capacitive sensors as a complex impedance which includes not only a capacitance C but also a parallel resistance R representing the dielectric losses. In an I/Q modulation circuit, the sensor is fed by an AC voltage, the currents I_c and I_r flowing though the capacitance and the resistance, being converted to voltage and sent to two phase sensitive detectors which compare this voltage with two 90° phased voltages. The two output voltages of the phase sensitive detectors are proportional to the values of C and R. This type of circuit is complex and the accuracy of the measurement is influenced by multiple factors, such as the accuracy of the phase shifters and the phase sensitive detectors [12]. Another type of circuit uses only the Q modulation in order to compensate the effect of the parallel resistance. A feedback controls the phase of the input signal in such a way it compensates the phase shift caused by the sensor resistance [13].

The use of the capacitance-to-voltage type of circuit is also to be considered; in this type of circuit only an accuracy of mV is required, while the capacitance-to-time type requires nanosecond accuracy and the capacitance-to-frequency type requires MHz accuracy. Among the capacitance-to-voltage types the transducer circuit is selected because of its simplicity and its high accuracy to measure capacitance changes compared to the I/Q type. The circuit proposed by Preethichandra in [10] is a capacitance-to-voltage type of circuit. A reference capacitance is used to compensate the effect of stray capacitance caused by the measurement circuit. It consists of two similar sub-circuits, one for the capacitance to measured C_x and one for the reference capacitance C_{ref}. Each sub-circuit consists of a current-to-voltage converter, a rectifier and a low-pass filter. The difference between the two output signals is calculated with a differential amplifier. The final output voltage V_{out} depends on C_x and C_{ref}. Since C_{ref} is kept constant, V_{out} depends in reality only on C_x. The bias voltages caused by stray capacitances in both sub-circuits are eliminated by the differential amplifier.

In this chapter, we study the possibility of using the capacitance-to-voltage type of circuit for the measurement of oil quality considering the effects of stray capacitances and dielectric losses.

2 Measurement circuit

A simplified schematic of the measurement circuit is presented in Fig. 1. The working principle is to inject a sinusoidal signal to the sensor and a reference capacitance; this creates currents flowing through these two capacitances. These currents are then converted to voltage, rectified and filtered. The filters outputs two DC signals depending on the sensor capacitance and the reference capacitance. The final output of the circuit, V_{out}, is the difference between both signals; V_{out} is maximum when C_x is minimum and it is theoretically 0 when $C_x = C_{ref}$. The circuit was adapted to oil measurement by adding a parallel resistance R_x to C_x and stray capacitances (C_{s1}, C_{s2} and C_{s3}) connected to C_x and C_{ref}.

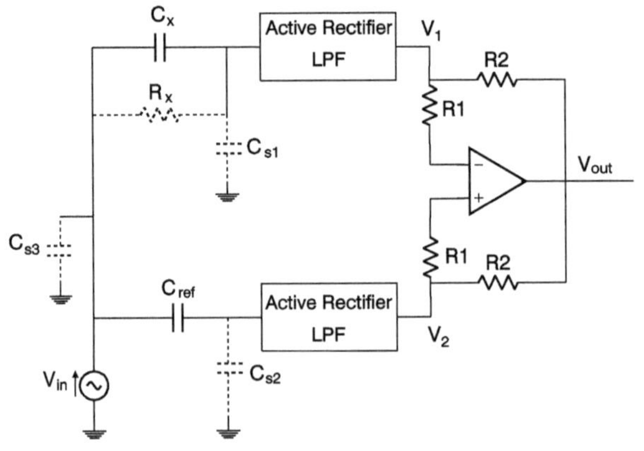

Fig. 1: Capacitance measurement circuit considering stray capacitances and dielectric losses.

2.1 Active rectifier

A precise full-wave rectifier is used in this circuit. This type of rectifier is considered more precise comparing with the half-wave rectifier, which keeps only half of the original input signal. In fact the full-wave rectifier reverses the negative part of the input signal and combines it with the positive part. The average (DC) output voltage is higher than for the half-wave rectifier and has much less ripples, producing a smoother output waveform.

The input amplifier (Fig. 2) controls the biasing of the diodes D1 and D2 to change the signal path based on the polarity of the input signal. Diode D1 provides the positive half wave to the output amplifier while the diode D2 provides the negative half. The negative part is inverted and summed to the positive half at the output amplifier.

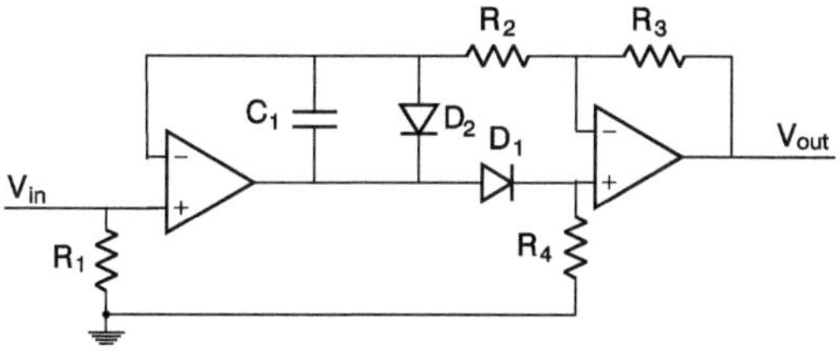

Fig. 2: Full-wave precision rectifier.

2.2 Low-pass filter

A low-pass filter is used to eliminate the ripples and high-frequency noises. A second-order filter based on Sallen–Key topology is used (Fig. 3). At low frequencies the capacitors C_1 and C_2 are considered open circuit and the signal is buffered directly to the output. At high frequencies C_1 and C_2 cannot be considered as open circuit and the input signal is shunted to the ground. The cut-off frequency of the filter can be calculated as follows:

$$f_c = \frac{1}{2\pi\sqrt{R_1R_2C_1C_2}}. \tag{1}$$

A low cut-off frequency can guarantee a smooth DC output, but this will decrease the quality factor Q and the gain of the filter.

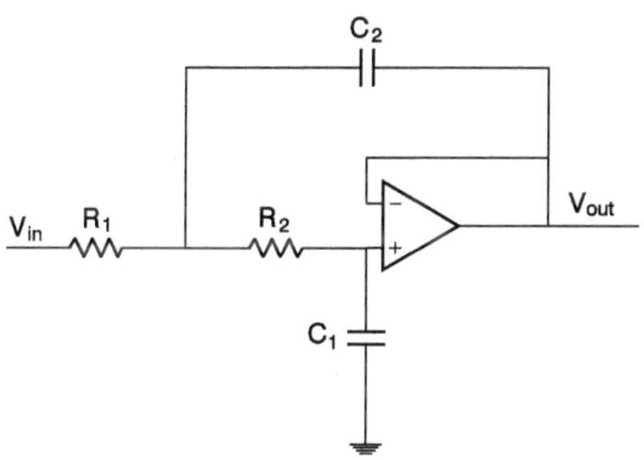

Fig. 3: Sallen–Key low-pass filter.

2.3 Differential amplifier

The function of the differential amplifier is to calculate the difference between the two output voltages of the low-pass filters, which represent the measured capacitance C_x and the reference capacitance C_{ref}.

The offset voltage caused by the non-ideality of the amplifiers should be low so that the output voltage is 0 when $C_x = C_{ref}$. It is possible to measure capacitances higher than the reference capacitance C_{ref} but only if the differential amplifier is supplied with ±Vcc. In the case the amplifier is supplied with +Vcc and the ground, it is only possible to measure capacitances lower than C_{ref}.

3 Simulation results

The circuit blocks were first separately implemented in the simulator in order to study their behaviors and select the values of their components. The full circuit is then implemented for further investigation. The current flowing through C_x is converted to voltage across the resistance R_1 of the rectifier (R_1 in Fig. 2). This voltage is then rectified and filtered by the full-wave rectifier and the low-pass filter; C_x and R_1 form a high-pass filter with a variable gain and cut-off frequency changing with the change of C_x and R_1. For frequencies higher than the cut-off frequency the gain of the filter is equal to 1; in this case it is not possible to detect the change of the capacitance because the amplitude remains constant, but for frequencies lower than the cut-off frequency the gain will depend on the value of C_x and the output amplitude will be changing with respect to the change of C_x. The gain can be calculated as follows:

$$\text{Gain} = \frac{R_1}{\sqrt{{R_1}^2 + \frac{1}{(\omega C_x)^2}}}. \tag{2}$$

The sensitivity of the filter gain to the change of the capacitance is studied. The input frequency f_{in} and the resistance R_1 are varied in the ranges of 100 Hz to 100 kHz and 1 kΩ to 10 MΩ. The maximum frequency is set at 100 kHz due to the limitation of the amplifiers. In fact the amplifiers to be used in such measurement circuits should have a low offset voltage and low bias current, though amplifiers which have a low offset voltage, in the range of µV, and bias current, in the range of pA, have very limited bandwidth in the range of 100–1000 kHz.

The maximum sensitivity (0.08 gain change for 100 pF C_x change) was found at different combinations of f_{in} and R_1. The input frequency f_{in} is selected to be 10 kHz (much smaller than 100 kHz), corresponding to $R_1 = 20$ kΩ.

In a first investigation, the circuit is simulated without stray capacitance and dielectric losses. The reference capacitance C_{ref} is kept constant at 500 pF and the voltages V_1, V_2 and V_{out} are recorded (Fig. 1). The voltage V_1 changes linearly with the change of the capacitance C_x, thus the voltage V_{out} changes also linearly with C_x (Fig. 4). The simulations were done using the ADA4522 amplifier, which has a maximum offset voltage equal to 5 µV and a maximum input bias current equal to 50 pA.

Ideally the output voltage equals to 0 V when $C_x = C_{ref}$, but due to the non-ideality of the amplifiers there is a voltage offset equal to 0.013 V at the output voltage. The relation between C_x and V_{out} is linear with a resolution of 2 mV/1 pF. A curve fitting

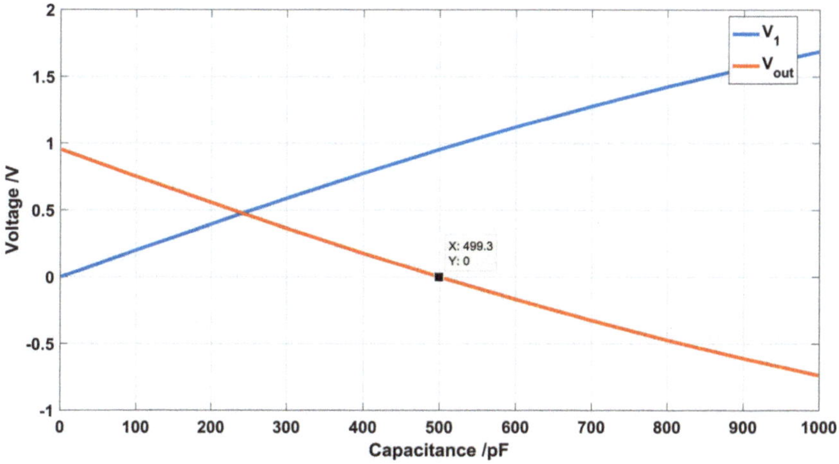

Fig. 4: Output voltage of the measurement circuit.

was used in order to find the relation between C_x and V_{out}, following the following equation:

$$C_{calc} = 0.495V_{out} + 492.3, \tag{3}$$

where C_{calc} is given in pF and V_{out} in mV.

To evaluate the goodness of the fit, the relative deviation was calculated using the following equations:

$$\text{Deviation} = |C_x - C_{calc}|, \tag{4}$$

$$\text{Relative deviation} = \frac{\text{Deviation} * 100}{C_x}. \tag{5}$$

The relative deviation decreases with the increase of the capacitance C_x; it has its maximum value 0.1 % at 1 pF and decreases to 0.02 % at 1000 pF, which means high linearity between C_x and V_{out} in the ideal case. In fact, the assessment of oil quality using capacitance sensors suffers from two main stray effects:

- the stray capacitances caused mainly by the sensor geometry and the interface between the sensor and the measurement circuit;
- the parallel resistance which represents the dielectric losses.

The effects of stray capacitances C_{s1}, C_{s2} and C_{s3} were first studied (Fig. 1). The values of C_{s1}, C_{s2} and C_{s3} were set at 10 pF, 50 pF and 100 pF.

A total of nine simulations have been made and the capacitance C_{calc}, the deviation and the relative deviation were calculated using equations (3), (4) and (5). The capacitance C_{s3} has minor effects on the measurement since it is connected in parallel to the voltage source, which has a low input resistance. The stray capacitances C_{s1}

and C_{s2} also have minor effects when their values are equal; this is because the error in voltage caused by C_{s1} is compensated by the error caused by C_{s2} ($V_{out} = V_2 - V_1$). The deviation in the former case was less than 1 % in all cases ($C_{s1} = C_{s2} = 10$, 50 or 100 pF). The deviation caused by $C_{s1} \neq C_{s2}$ is much higher than in the previous cases. It increases to a maximum of 8.5 % at 100 pF, decreases significantly around the reference capacitance and increases again up to 2.5 % at 1000 pF (Fig. 5). The importance of the reference sub-circuit is to compensate the stray effects. Using the reference capacitance it is possible to guarantee a maximum deviation of 1 % in the range of 400–600 pF in the presence of 100 pF stray capacitance.

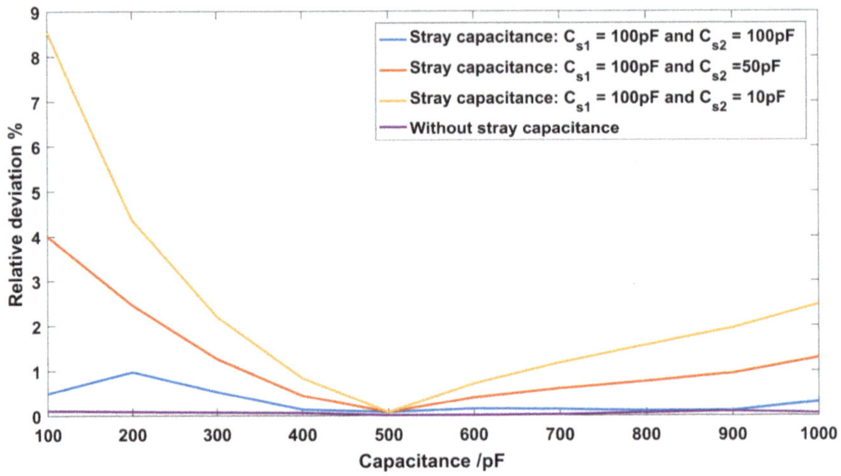

Fig. 5: Relative deviation caused by the stray capacitances.

The second parameter to be considered is the parallel resistance to the sensor capacitance; this resistance R_x represents the dielectric losses of the dielectric medium under investigation. The value of this parallel resistance varies depending on the sensor geometry and the type of the medium. In the case of edible oils the value of R_x varies depending on the type and quality of the oil. A typical value of R_x is in the MΩ range. For this purpose, a parallel resistance is added to the capacitance C_x as illustrated in Fig. 1. The circuit is simulated with and without stray capacitances for resistance R_x in the range of 1–100 MΩ.

With the presence of the parallel resistance, C_x, R_1 and R_x form a high-pass filter expecting that the gain is constant at low frequencies (Fig. 6). We have

$$H(w) = \frac{R_1}{R_1 + \frac{R_x}{1+j\omega C_x R_x}} \tag{6}$$

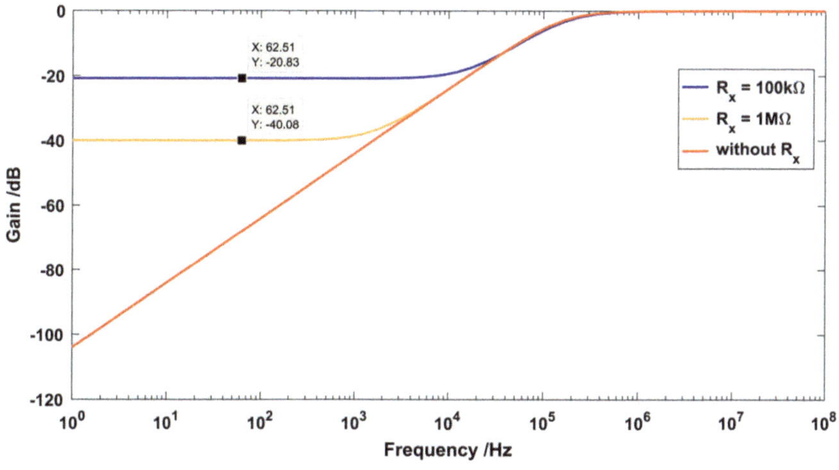

Fig. 6: Magnitude Bode plot of the filter formed by C_x, R_1 and R_x.

The new transfer function depends on C_x, R_1 and R_x (equation (6)); at low frequencies ($\omega = 0$), $H(\omega) = \frac{R_1}{R_1+R_x}$, while at high frequencies ($\omega = \infty$), $H(\omega) \approx 1$. The gain at low frequencies equals $\frac{R_1}{R_1+R_x}$, which is different from the case without R_x.

In the case $R_x = 100\,\text{k}\Omega$, the relative deviation is higher than 100 % and it was not possible to calculate C_x using equation (3). In this case the current flows through R_x only. This makes the circuit behave like an inverting amplifier with a constant gain independent of the value of C_x. For higher values of R_x the relative deviation decreases; at $R_x = 10\,\text{M}\Omega$ it has its maximum (1 %) at 100 pF while it has its minimum (0.6 %) for values near the reference capacitance (Fig. 7). In the case where both stray capacitance

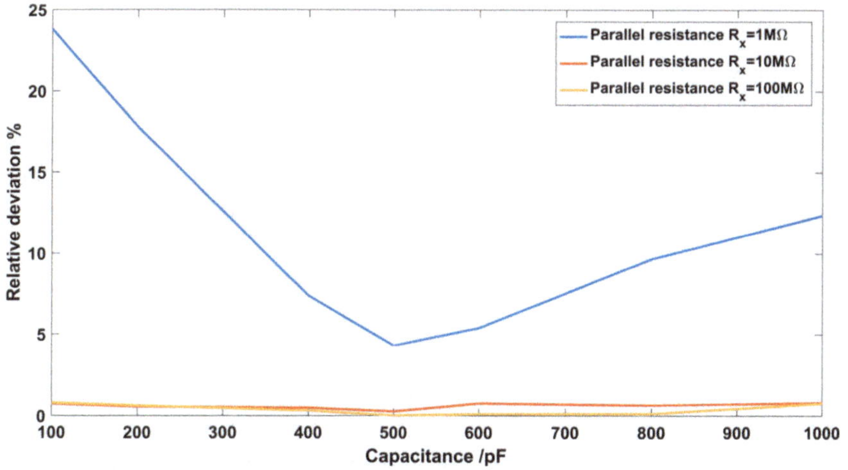

Fig. 7: Relative deviation caused by the parallel resistance.

and parallel resistance are added to the circuit, the deviation effect is more by the parallel resistance; in fact, at a stray capacitance of 10 pF and when $R_x = 10\,\text{M}\Omega$, the maximum deviation was 2 % at 100 pF and it decreases for higher capacitance values.

4 Experimental results

The circuit was designed and implemented on a printed board for experimental investigation. The amplifier ADA4522 was used in the circuit and it was selected because of its low offset voltage and bias current. Two amplifiers were used so that the two rectifiers and the two low-pass filters (measurement and reference branches) could be implemented in a way the error caused by stray capacitances of the measurement circuit would be similar and hence compensated in the differential amplifier. The diode BAS16VY was used in the rectifier circuit; it was selected because of its very small reserve recovery time (minimum time to switch from passing to blocking modes and vice versa).

The measurement circuit is excited by a sine wave signal with a 2-Vpp amplitude at a frequency of 10 kHz. The waveform generator AD9933 was used because of its low cost, simple interface and stable output frequency and amplitude, especially at low frequencies. A micro-controller with an accurate ADC (12 bits resolution and 250 kS/S sampling rate) was used to measure the output voltage.

A set of 30 capacitances in the range of 1–700 pF were accurately measured with a precision impedance analyzer for the calibration of the measurement circuit. The reference capacitance is chosen to be 500 pF; this limits the measurement range to 500 pF, so that no negative voltage will be in the output of the circuit, which can damage the ADC of the micro-controller.

An offset voltage of 60 mV was measured when $C_x = C_{\text{ref}} = 500\,\text{pF}$; this can be caused by the measurement circuit and the error of the analog-to-digital conversion. It was not possible to measure capacitances lower than 10 pF because the output voltage was not sensitive to the change of capacitance in this range, while it linearly decreases with the increase of capacitance in the range of 10–500 pF. The practical relation between the output voltage and input capacitance was developed using the curve fitting tool, yielding

$$\text{Capacitance} = 0.5304V_{\text{out}} + 468.3. \tag{7}$$

The maximum relative deviation is 30 % for capacitances smaller than 70 pF while it decreases to 2% at 100 pF and decreases further when the input capacitance increases. A minimum change of 3 pF was accurately detected, corresponding to a change of 6 mV in the output voltage.

5 Conclusion

The proposed measurement circuit is suitable for the measurement of small capacitance changes in the pF range in the capacitance range from 10–500 pF. It was possible to reduce the effect of stray capacitances using the reference principle. The measurement accuracy increases for capacitance values near the reference capacitance and hence it is advised to used a reference capacitance equal to the nominal capacitance of the sensor. The dielectric losses have major effects on the measurement; in fact this circuit is not suitable for the measurement of high-loss materials, though it is possible to use this circuit for the assessment of oil quality since it has low dielectric losses.

Bibliography

[1] H. Lizhi, K. Toyoda, and I. Ihara, "Dielectric properties of edible oils and fatty acids as a function of frequency, temperature, moisture and composition", Journal of Food Engineering, vol. 88, no. 2, pp. 151–158, 2008.

[2] H. Lizhi, K. Toyoda, and I. Ihara, "Discrimination of olive oil adulterated with vegetable oils using dielectric spectroscopy". Journal of Food Engineering, vol. 96, no. 2, pp. 67–171, 2010.

[3] A. Fendri, O. Kanoun, and H. Ghariani, "Dielectric spectroscopy for assessment of water content in edible oils", In: Systems, Signal & Devices (SSD), 14th International Multi-Conference on Sensor System & Devices. IEEE, 2017.

[4] A. Y. Khaled and S. A. Aziz Rokhani, "Capacitive sensor probe to assess frying oil degradation". Information Processing in Agriculture, vol. 2, no. 2, pp. 142–148, 2015.

[5] A. Heidary and G. C. Meijer, "An integrated interface circuit with a capacitance-to-voltage converter as front-end for grounded capacitive sensors", Measurement Science & Technology, 2008.

[6] R. Nojdelov and S. Nihtianov, "Capacitive-sensor interface with high accuracy and stability". IEEE Transactions on Instrumentation and Measurement, 1633–1639, 2000.

[7] M. Z. Aslam and T. B. Tang, "A high resolution capacitive sensing system for the measurement of water content in crude oil", pp. 11351–11361.

[8] R. A. Brookhuis, T. S. Lammerink, and R. J. Wiegerink, "Differential capacitive sensing circuit for a multi-electrode capacitive force sensor", Sensors and Actuators A, Physical, pp. 168–179, 2015.

[9] M. D. Benadda, J. C. Carru, and C. Druon, "A measuring device for the determination of the electric permittivity of materials in the frequency range 0.1–300 MHz", Journal of Physics E, Scientific Instruments, 1982.

[10] D. M. Preethichandra and K. Shida, "A simple interface circuit to measure very small capacitance changes in capacitive sensors". IEEE Transactions on Instrumentation and Measurement, pp. 1583–1586, 2001.

[11] S. M. Huang, C. G. Xie, A. L. Stott, R. G. Green, and M. S. Beck, "A capacitance-based solids concentration transducer with high immunity to interference from the electrostatic charge generated in solids/air two-component flows", Transactions of the Institute of Measurement and Control, pp. 98–102, 1988.

[12] D. Marioli, E. Sardini, and A. Taroni, "High-accuracy measurement techniques for capacitance transducers", Measurement Science & Technology, 1993.

[13] A. Pinter and I. Dénes, "Interface circuit for measuring small capacitance changes in sensor networks", IET Science, Measurement & Technology, pp. 570–578, 2015.

Part III: **Sensors**

Christian Weber, Markus Tahedl, and Olfa Kanoun

Modeling for improved performance of non-contacting capacitive sensors for detecting aqueous solutions

Abstract: In industrial applications, non-contacting capacitive sensors are used to detect conductive fluids in containers. During use, the material to be detected can stick to the inside of the container, leading to measurement deviations. Using analytical modeling, it has been found that an overall resistance can distinguish between the conductive film and the actual fill level. Impedance analyzers have limitations with respect to their maximum measurement frequency of approximately 100 MHz. This bandwidth is not sufficient for characterizing highly conductive media. In order to overcome this limitation, improved signal processing is proposed to determine the overall resistance. In this paper, two methods of parameter extraction using the total measured bandwidth and a limited frequency range are compared. Results show that the overall resistance can be extracted using only a limited frequency range.

Keywords: impedance sensors, capacitive sensors, parameter extraction, modeling, impedance spectroscopy

1 Introduction

Limit levels of a container filled with a conductive fluid can be detected using capacitive sensors. For easy installation, the sensors are mounted outside of the non-conductive container. Conductive films inside of the container influence the sensor signal and can produce a false positive detection of the medium [1].

Fig. 1 shows a capacitive limit level switch mounted outside a container. Since in many applications the exact geometry, i. e., container material and wall thickness, and material properties of the medium are unknown, the sensors have to be calibrated before use. This is often done by adjusting the switch point using a potentiometer. Simple capacitive proximity switches measure only the absolute impedance between their sensing electrode and ground potential [2, 3]. Depending on the measurement frequency, wall thickness and conductivity of the fluid, the measured value of such sensors is the same for both the actual fill level and a conductive film.

Christian Weber, ifm electronic gmbh, Essen, Germany, e-mail: christian.weber@ifm.com
Markus Tahedl, Oertli Instrumente AG, Berneck, Switzerland
Olfa Kanoun, Chair for Measurement and Sensor Technology, Chemnitz University of Technology, Reichenhainer Strasse 70, 09126 Chemnitz, Germany

https://doi.org/10.1515/9783110558920-011

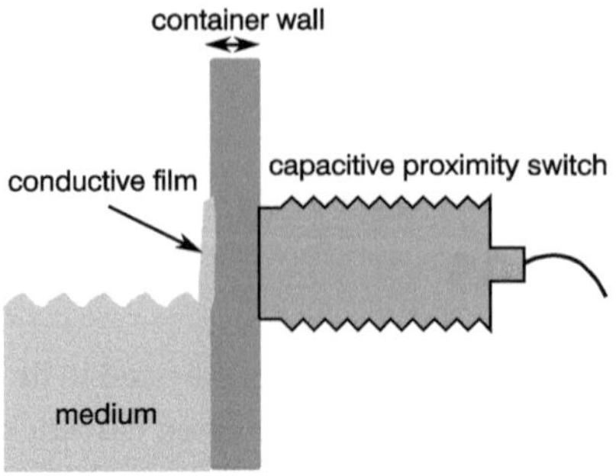

Fig. 1: Capacitive proximity switch mounted outside a container for detecting the limit level of the medium. Conductive films may stick to the inside of the container wall.

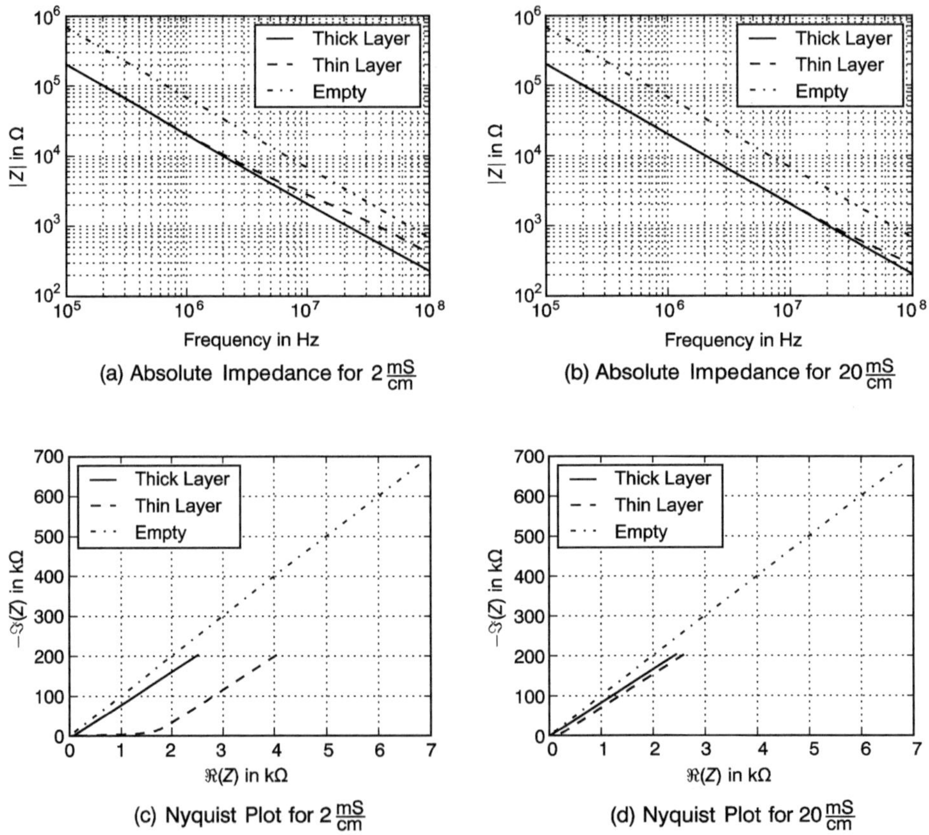

Fig. 2: Example plots to illustrate the difference between a thin layer and the actual full container for two different conductivities.

To illustrate the problem, Fig. 2 shows example spectra for a full container and a thin conductive film of a conductive aqueous solution for two conductivities 2 mS/cm and 20 mS/cm. It is shown in Fig. 2 (a) that a thin conductive film cannot be distinguished from the actual fill level below a frequency of 2 MHz. A difference between a thin and a thick layer can only be observed at frequencies above 2 MHz. This effect is even worse at higher conductivities, as can be seen in Fig. 2 (b), where the difference between thin and thick layers is only observable at frequencies higher than 20 MHz. The necessary measurement frequency increases approximately proportionally with the conductivity of the material to be detected.

Fig. 3 shows the difference of the absolute impedance, $\Delta|Z| = |Z_{\text{thin}}| - |Z_{\text{thick}}|$, between a thin layer and a thick layer for several conductivities. It can be seen in the figure that the maximum difference decreases with increasing conductivity. Furthermore, the measurement frequency at which the maximum difference occurs also increases with conductivity.

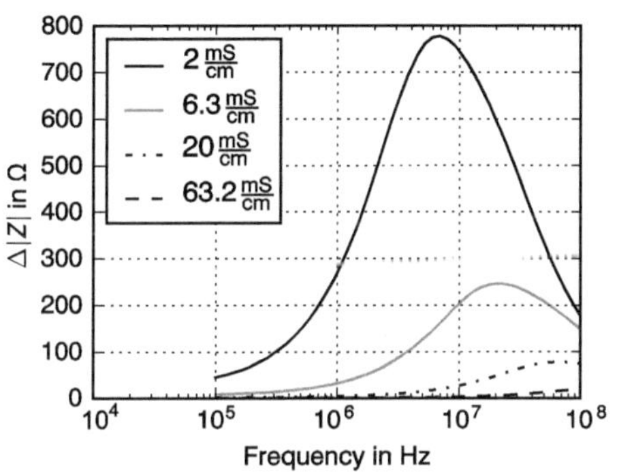

Fig. 3: Absolute difference $\Delta|Z| = |Z_{\text{thin}}| - |Z_{\text{thick}}|$ between a thin layer and a thick layer of the medium. Increasing the frequency reduces the observed difference and increases the necessary measurement frequency.

Since measurement devices are limited in terms of measurement frequency, another method must be found. Figs. 2 (c) and 2 (d) show complex Nyquist plots of the same dataset. It is possible to distinguish thin films and actual limit levels using impedance spectroscopy and associated signal processing techniques. The low-frequency part, where the Nyquist plot forms a straight line, can be shifted to the right by a decrease in layer thickness or a decrease in conductivity. Therefore, for a given conductivity, different layer thicknesses can be distinguished by inspection of the complex impedance plot even though there is no difference visible in the absolute value of the impedance.

Using analytical modeling, it was found that the overall resistances of a capacitively coupled aqueous solution can be used to distinguish between the conductive film and the actual limit level [1, 4]. To validate the proposed criterion experimentally, a measurement setup is needed to create thin layers of aqueous solutions.

This setup has already been described in [5]. However, the fitting methods are reported to have some limitations. Relative fitting errors of the proposed fitting procedures are quite large, up to 20 % in the real part of the measured spectrum.

In this contribution, improved fitting procedures are proposed to increase the quality of fit, which has been improved to below 0.5 % and 7 % for the imaginary and the real part, respectively. For very high conductivities, e. g., aqueous KCl solutions with a conductivity of up to 208 mS/cm [6, 7], a linear fitting procedure [1] is introduced. In the next sections, we introduce and compare two fitting procedures.

For low solution conductivity, the parameters of a behavioral equivalent circuit model are extracted using differential evolution. For high solution conductivity, low-frequency data are fitted to a first-order polynomial and the overall resistance is determined.

2 Extraction of the overall resistance by non-linear fitting

To adequately model the behavior of measured and simulated impedance spectra, the electrical equivalent circuit (EEC) shown in Fig. 4 can be used.

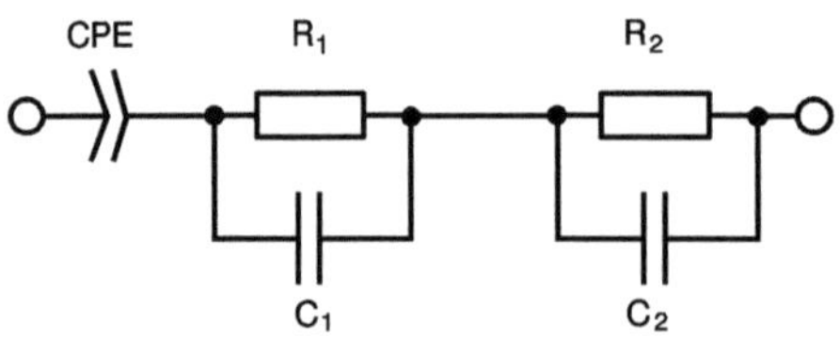

Fig. 4: Behavioral model for the complex impedance of capacitively coupled conductive media.

Since the electrode is mounted outside a non-conducting container, the capacitive coupling into the material under test (MUT) can be modeled using a constant phase element (CPE). Its impedance can be parameterized as follows [8]:

$$Z_{\text{CPE}} = \frac{1}{K} \cdot (j\omega)^n, \tag{1}$$

where j is the imaginary unit and K denotes the amplitude of the CPE; n can be any real value between −1 and 1 and is a measure for the dielectric loss factor D. We have

$$D = \frac{\Re(Z)}{\Im(Z)}. \tag{2}$$

If $n = 0$, the CPE becomes an Ohmic resistance $R = \frac{1}{K}$. For $n = -1$, the equation yields an ideal capacitance $C = K$; $n = 1$ yields an inductance $L = \frac{1}{K}$.

To link n and D, we rewrite equation (1) using [8] and we obtain

$$Z_{\text{CPE}} = \frac{\omega^n}{K} \cdot \left(\cos\left(n\frac{\pi}{2}\right) + j \cdot \sin\left(n\frac{\pi}{2}\right)\right). \tag{3}$$

Substituting into equation (2) yields

$$D = \frac{\cos(n\frac{\pi}{2})}{\sin(n\frac{\pi}{2})} = \frac{1}{\tan(n\frac{\pi}{2})}. \tag{4}$$

The equations for the CPE phase angle φ and quality factor Q are then

$$\varphi = n\frac{\pi}{2}, \tag{5}$$

$$Q = \frac{1}{D} = \tan\left(n\frac{\pi}{2}\right). \tag{6}$$

If an alternating electric potential is applied to an electrode, the equation of continuity of the electric current density states that the current can be separated into a dielectric and a conductive part [9]. Thus, the behavior of a conductive aqueous solution can be described by a series of resistor–capacitor (RC) elements. The impedance of a single RC element is

$$Z_{\mathrm{RC}} = \frac{R}{1 + j\omega RC}. \tag{7}$$

Substituting $\omega_{\mathrm{g}} = \frac{1}{\mathrm{RC}}$ yields

$$Z_{\mathrm{RC}} = \frac{R}{1 + j\frac{\omega}{\omega_{\mathrm{g}}}}. \tag{8}$$

Thus, the model impedance Z_{mod} is

$$Z_{\mathrm{mod}} = \frac{1}{K} \cdot (j\omega)^n + \frac{R_1}{1 + j\frac{\omega}{\omega_{\mathrm{g},1}}} + \frac{R_2}{1 + j\frac{\omega}{\omega_{\mathrm{g},2}}}. \tag{9}$$

To extract the model parameters of the EEC shown in Fig. 4, a differential evolution algorithm as proposed in [5] was used. Evolutionary algorithms benefit from a large search space and robust convergence behavior [10, 11]. Compared to gradient methods, which tend to get trapped in local minima of the cost function, especially at increased noise levels, stochastic methods are more robust because no gradients are calculated [12], though it is worth mentioning that the noise level of an optimization problem also depends on the cost function employed [13].

The first two points are of special interest here, since the impedance spectra of the EEC in question require a high-frequency bandwidth for all parameters to be extracted [5].

For the fitting procedure to succeed, a suitable cost function must first be defined. Since the measured data represent a capacitively coupled conductive fluid, the phase angle of the measured impedance is always relatively close to −90°. It is therefore not recommended to use the absolute difference between measured and modeled data as a cost function since it is mainly sensitive to changes in the imaginary part of the spectrum. Instead, we propose to adopt a parametric cost function c as proposed by [14] to describe the modeling error for N data points and M parameters, i. e.,

$$c = \frac{1}{N - M} \sum_{\mathrm{i}=1}^{N} \left(\frac{(\Re(Z_{\mathrm{i,meas}}) - \Re(Z_{\mathrm{i,mod}}))^2}{\Re(Z_{\mathrm{i,meas}})^2} + \frac{(\Im(Z_{\mathrm{i,meas}}) - \Im(Z_{\mathrm{i,mod}}))^2}{\Im(Z_{\mathrm{i,meas}})^2} \right). \tag{10}$$

This cost function describes the sum of relative quadratic residuals of the measured data. This way, the real and imaginary residuals at all frequencies contribute equally to the cost function, which allows for a better extraction of the unknown parameters. When using differential evolution, extracted parameters can depend on the parameter boundaries specified. Since model parameters must be found over a large frequency bandwidth and the algorithm should work on as many different spectra as possible, parameter boundaries also become very broad, ranging over several decades. Thus, we propose to employ logarithmic compression to shrink the search space, i. e.,

$$R_{\log} = \log_{10}\left(\frac{R}{1\Omega}\right). \tag{11}$$

For example, if the search space for a single resistance ranges from $1\,\Omega$ to $1\,\mathrm{M}\Omega$, one substitutes both of these values into equation (11), which yields unitless values of 0 and 6, respectively.

At low frequencies, the real and imaginary parts of an impedance spectrum of a capacitively coupled aqueous solution have a high sensitivity with respect to the CPE parameters [5]. Small perturbations in these parameters will cause a large fitting error. To guarantee a reliable extraction of the CPE parameters, we suggest to first estimate the CPE parameters from the measured data and then choose the initial population boundaries to be within ±50 % of the estimated values. The CPE values can be estimated by linearly fitting the low-frequency part of the spectrum to a first-order polynomial, i. e.,

$$\Re(Z) = D_{\mathrm{E}} \cdot \Im(Z) + R_{\mathrm{O}}. \tag{12}$$

Fig. 5 shows an illustration of the linear fitting procedure. For a successful estimation, the part of the Nyquist plot where the impedance curve forms a straight line must be used. The imaginary part of the spectrum is used as an independent variable; R_{O} is the intercept point with the real axis at $\Im(Z) = 0$. The loss factor D_{E} represents the slope of the low-frequency line.

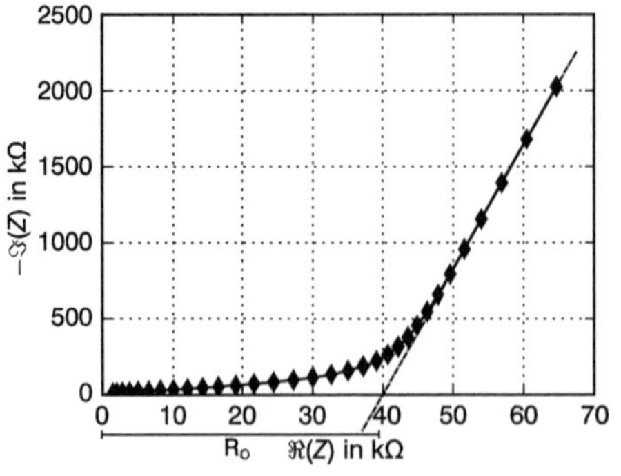

Fig. 5: Illustration of the estimation of overall resistance and CPE parameters.

Using equation (5), the exponent n_{ab} of CPE can be approximated; K can be approximated by rearranging equation (1), i. e.,

$$K_{ab} = \frac{1}{Z(\omega_1) - R_0} \cdot (j\omega_1)^{n_{ab}}, \tag{13}$$

where $Z(\omega_1)$ is the measured impedance at the lowest frequency. With the estimated CPE parameters and logarithmic compression for all RC elements, the initial population is computed using latin hypercube sampling [15]. After convergence of the differential evolution optimizer, the implementation [15] employed in this work can optionally refine the stochastic result with a direct search optimizer [16, 17]. If the direct search optimizer yields, in terms of cost, a better result than the evolution algorithm, its solution is replaced. If not, the result of the differential evolution is returned.

2.1 Evaluation of the proposed modifications

To evaluate the proposed method, a noiseless test dataset using the EEC shown in Fig. 4 is generated in a frequency range from 1 kHz to 100 MHz with the nominal values shown in Table 1. The modified extraction method with logarithmic compression and approximation of the CPE parameters was compared against a classical differential evolution approach. For both methods the algorithm was set up with the same population size of 80 individuals, 10^4 maximum iterations and the same convergence criterion described in [15]. The algorithm terminates if

$$\frac{\text{stdev (population)}}{\text{mean (population)}} < 10^{-3} \tag{14}$$

is true, where stdev (population) denotes the standard deviation of the population. All other parameters of the optimizer were left at the standard values provided by its implementation in the Python module Scipy 0.18 [15].

In the implementation used in this study, C_1 and C_2 are also treated as CPEs to make the algorithm easily expandable to more complex models, though their exponents n are chosen in such a way that their loss factor does not exceed 1.6×10^{-12}.

Table 1: Result of the parameter reconstruction.

Parameter	Nominal value	Modified reconstruction	Classical reconstruction
n	−0.9923	−0.9923	−0.9923
K in $\frac{1}{\Omega}(\frac{\text{rad}}{\text{s}})^n \cdot 10^{-12}$	4	4	4
C_1 in pF	150	149.99	10^4
C_2 in pF	14.5	14.5	13.2
R_1	500.00	500.00	999633.70
R_2	6600	6599.99	7086.71
Cost function c	0	2.61×10^{-20}	1.01×10^{-6}

The boundary values for the search space of the resistances is $[0, 6]$ (logarithmic) for the modified method and $[1\Omega, 1\,\mathrm{M}\Omega]$ for the classical method. For capacitances, intervals of $[-13, -8]$ (logarithmic) and $[10^{-13}\mathrm{F}, 10^{-8}\mathrm{F}]$ were chosen. For the coupling CPE, the modified method employs boundary values of $[0.5{\times}K_{\mathrm{ab}}, 1.5{\times}K_{\mathrm{ab}}]$ and $[-1, 0.9{\times} n_{\mathrm{ab}}]$, while the classical method uses $[10^{-13}\frac{1}{\Omega}(\frac{\mathrm{rad}}{\mathrm{s}})^n, 10^{-8}\frac{1}{\Omega}(\frac{\mathrm{rad}}{\mathrm{s}})^n]$ for the CPE amplitude and $[-1, -0.8]$ for the CPE exponent n. Implementation was done using [15, 18].

Table 1 shows that a reconstruction of the nominal parameters is possible with the proposed method. With the classical method, the algorithm seems to converge to a local minimum, as can be seen in the rightmost column of the table.

Furthermore, the result of the classical method is not always reproducible. Multiple optimization runs yield different parameter vectors with cost function values, which in some cases are as high as $c \approx 0.3$.

For the classical method to converge to the correct solution, the parameter boundaries have to be narrowed by two decades to $[10^{-12}\mathrm{F}, 10^{-9}\mathrm{F}]$ for the capacitances and $[1\Omega, 10\,\mathrm{k}\Omega]$ for the resistances. This shows that using logarithmic compression increases conversion reliability and user convenience.

However, at the chosen population size of 80 individuals, using linear fitting to estimate the CPE parameters does not seem to be necessary as long as logarithmic compression is used. The evolution algorithm also converges on the nominal parameter values even if the parameter boundaries for the coupling CPE are set to be $[10^{-13}\frac{1}{\Omega}(\frac{\mathrm{rad}}{\mathrm{s}})^n, 10^{-8}\frac{1}{\Omega}(\frac{\mathrm{rad}}{\mathrm{s}})^n]$ and $[-1, -0.8]$.

Although the linear parameter estimation does not yield significant improvement to the non-linear extraction, it is still useful to characterize solutions where the measured spectrum does not contain enough information about all RC elements.

3 Extraction of the overall resistance by linear fitting

Because of the limited measurement bandwidth, it is not always possible to extract all time constants of a given measured spectrum. However, one can still obtain information about the overall resistance and the coupling CPE. Instead of fitting the impedance spectrum over its full bandwidth, one can use equation (12), which is used to extract CPE parameters before a non-linear regression. It has already been shown experimentally in a previous work that this parameter is related to layer thickness and conductivity of the fluid [1]. Using a behavioral model with a CPE and one RC element with time constant $\tau = R_1C_1$, it can be shown that R_0 is an approximation of R_1. Most symbolic calculations have been carried out using [19]. Separation of real and imaginary part of the impedance yields

$$\Re(Z) = \frac{1}{K(C_1^2R_1^2\omega^2+1)}\left(C_1^2R_1^2\omega^{n+2}\cos\left(\frac{\pi n}{2}\right) + KR_1 + \omega^n\cos\left(\frac{\pi n}{2}\right)\right), \tag{15}$$

$$\Im(Z) = \frac{1}{K(C_1^2R_1^2\omega^2+1)}\left(C_1^2R_1^2\omega^{n+2}\sin\left(\frac{\pi n}{2}\right) - C_1KR_1^2\omega + \omega^n\sin\left(\frac{\pi n}{2}\right)\right). \tag{16}$$

In the frequency region where the Nyquist plot forms a straight line, two frequencies ω_1 and ω_2 are selected. The line slope D_E can be approximated as follows:

$$D_E(\omega_1,\omega_2) = \frac{\Re(Z(\omega_1)) - \Re(Z(\omega_2))}{\Im(Z(\omega_1)) - \Im(Z(\omega_2))}. \tag{17}$$

By substituting equation (17) into equation (12) and solving for R_O we get

$$R_O(\omega_1,\omega_2) = \Re(Z(\omega_1)) - D_E(\omega_1,\omega_2)\cdot\Im(Z(\omega_1)). \tag{18}$$

We now let the difference between the two frequencies approach zero by calculating the limit, i. e.,

$$R_O(\omega_2) = \lim_{\omega_1\to\omega_2} R_O(\omega_1,\omega_2), \tag{19}$$

and we get

$$R_O(\omega_2) = \frac{R_1\left(\frac{C_1R_1\omega_2}{\tan\left(\frac{\pi n}{2}\right)}+1\right)}{C_1^2R_1^2\omega_2^2+1}. \tag{20}$$

For simplicity, we also substitute $R_1C_1 = \frac{1}{\omega_g}$, so we obtain

$$R_O(\omega_2) = \frac{R_1}{\frac{\omega_2^2}{\omega_g^2}+1}\left(\frac{\omega_2}{\omega_g\tan\left(\frac{\pi n}{2}\right)}+1\right). \tag{21}$$

Equation (21) has a pole for $n = 0$. However, this pole is not relevant in the application discussed here, because the CPE becomes a pure resistance in that case, thus removing the capacitive coupling from the system; R_O is a good approximation of R_1, if $\omega_g \gg \omega_2$ holds. In this case, the terms $\frac{\omega_2^2}{\omega_g^2}$ and $\frac{\omega_2}{\omega_g\tan\left(\frac{\pi n}{2}\right)}$ both approach zero and the equation becomes $R_O = R_1$. Typically, $\omega_g \gg \omega_2$ is fulfilled for highly conductive media since the low resistances cause high cut-off frequencies.

It can also be shown that D_E is a good approximation for the impedance loss factor D. The loss factor of an EEC containing one RC element and one CPE is

$$D = \frac{\frac{KR_1}{\frac{\omega^2}{\omega_g^2}+1} + \frac{\omega^2\omega^n\cos\left(\frac{\pi n}{2}\right)}{\omega_g^2\left(\frac{\omega^2}{\omega_g^2}+1\right)} + \frac{\omega^n\cos\left(\frac{\pi n}{2}\right)}{\frac{\omega^2}{\omega_g^2}+1}}{-\frac{KR_1\omega}{\omega_g\left(\frac{\omega^2}{\omega_g^2}+1\right)} + \frac{\omega^2\omega^n\sin\left(\frac{\pi n}{2}\right)}{\omega_g^2\left(\frac{\omega^2}{\omega_g^2}+1\right)} + \frac{\omega^n\sin\left(\frac{\pi n}{2}\right)}{\frac{\omega^2}{\omega_g^2}+1}}. \tag{22}$$

If $\omega_g \gg \omega$ holds, we can calculate the limit, i. e.,

$$D_E = \lim_{\frac{\omega}{\omega_g}\to 0} D = \frac{\omega^{-n}}{\sin\left(\frac{\pi n}{2}\right)}\left(KR_1 + \omega^n\cos\left(\frac{\pi n}{2}\right)\right), \tag{23}$$

which we can write as

$$D_E = \frac{\omega^{-n} K R_1}{\sin(\frac{n\pi}{2})} + \frac{\cos(\frac{n\pi}{2})}{\sin(\frac{n\pi}{2})}. \tag{24}$$

The denominator of the first additive term of equation (24) is close to −1 since n is also close to −1 for a capacitively coupled medium. Also, $\omega^{-n} \approx \omega$ holds for the same reason. However, since K is typically small, approximately $10^{-12} \frac{1}{\Omega}(\frac{\text{rad}}{\text{s}})^n$ to $10^{-10} \frac{1}{\Omega}(\frac{\text{rad}}{\text{s}})^n$, it can be stated that for small values of R_1 the first additive term approaches zero. Thus, D_E approaches D.

Equations (21) and (24) show that a linear fitting routine can be used to extract the two CPE parameters and the overall resistance R_0, which is closely related to the actual resistance of the fluid. Therefore, even though the bandwidth of the measurement setup is limited, it can still be used to characterize highly conductive media.

4 Experimental investigations

The proposed fitting methods are now validated using measured data. The media and conductivities used are listed in Table 2.

Table 2: Conductivity of the measured media.

Medium	Conductivity σ in mS/cm
Tap water	0.5
KCl #1	1.9
KCl #2	10.6
KCl #3	43.1
KCl #4	92.5
KCl #5	193.1

Fig. 6 shows a concept drawing of the proposed measurement setup. It consists of an outer and an inner cylinder fabricated from plastic. The inner cylinder is hollowed out to minimize its influence of the measured impedance. Its height can be adjusted to create a defined gap at the bottom of the setup, where the measurement electrode is mounted. The height of this gap will be referred to as layer thickness. The aqueous solution to be measured is inserted between the outer and the inner cylinder. A detailed description is available in [2]. Impedance spectra were recorded at 801 frequency points, the maximum amount for a single sweep, in a range from 4 kHz to 110 MHz.

Measurement noise of the impedance measurement instrument depends on various factors, i. e., the connected device under test, the selected frequency range and

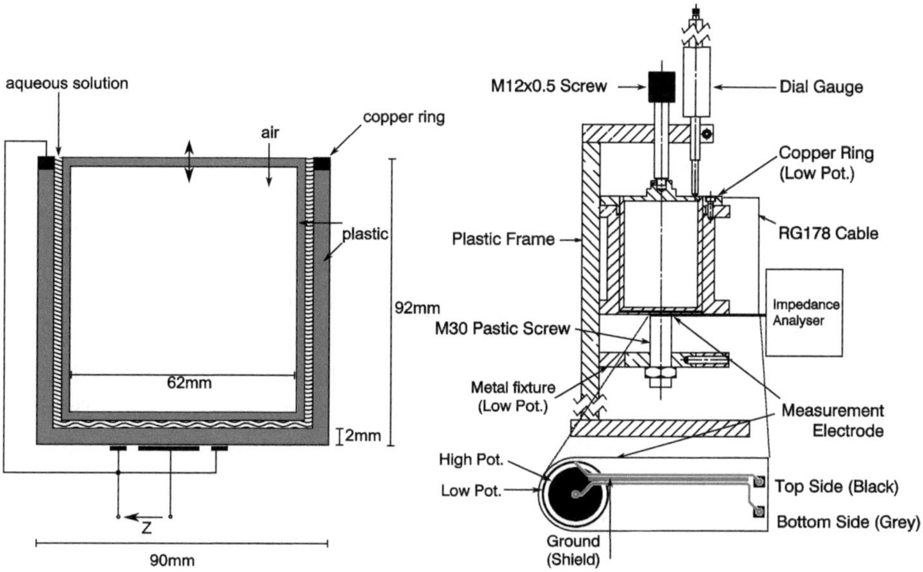

(a) Concept drawing

(b) Detailed drawing including measurement electrode layout

Fig. 6: Measurement setup drawings [5]. The setup is connected to an impedance measurement instrument, which provides high, low and ground potential terminals.

the test signal level [3]. Measurement noise for capacitively coupled resistances characterized in this work was approximated by connecting an inter-digital capacitor on an FR4 printed circuit board with CPE values $K = 5.7 \times 10^{-12} \frac{1}{\Omega} (\frac{\mathrm{rad}}{\mathrm{s}})^n$ and $n = -0.99$ connected in series to a 150-Ω resistor. A total of 600 spectra with 50 points each in a frequency range from 1 MHz to 5 MHz were recorded. The maximum relative standard deviations for the absolute impedance and its phase angle are 0.07 % and 0.007 %, respectively.

Measured data are smoothed using a linear fitting algorithm [21] with the modifications described in [1, 22, 23]. For all models, the relative residuals for real Δ_{re} and imaginary Δ_{im} part of the impedance are calculated as follows:

$$\Delta_{\mathrm{re}} = \frac{\Re(Z_{\mathrm{meas}}) - \Re(Z_{\mathrm{mod}})}{\Re(Z_{\mathrm{meas}})} \cdot 100\,\%, \tag{25}$$

$$\Delta_{\mathrm{im}} = \frac{\Im(Z_{\mathrm{meas}}) - \Im(Z_{\mathrm{mod}})}{\Im(Z_{\mathrm{meas}})} \cdot 100\,\%, \tag{26}$$

where Z_{meas} denotes the measured impedance while Z_{mod} is the modeled impedance. A measured impedance is excluded from the spectrum, if its relative residual in either the real or the imaginary part exceeds 0.5 %. However, even though the fitting procedure itself is optimized, the selection of error thresholds and frequency bandwidth must still be done by the user. Therefore, different media can have different frequency ranges.

4.1 Extraction by non-linear fitting

The filtered spectra for tap water and KCl #1 are then used to validate the improved fitting procedure. All other media are employed to validate the linear fitting method, since sensitivity analyses have shown [5] that the measurement bandwidth for high conductivities is insufficient to extract all parameters of the EEC. Fig. 7 shows the relative residuals for tap water and KCl #1. A series inductance L_s has been added to the EEC to model the influence of cables connected to the measurement setup. This significantly reduces the relative residual in the imaginary part for frequencies greater than 10 MHz. The relative error of the imaginary part, shown in Figs. 7 (d) and 7 (c), is now below 0.5 %. The extracted value does not change over varying layer thickness

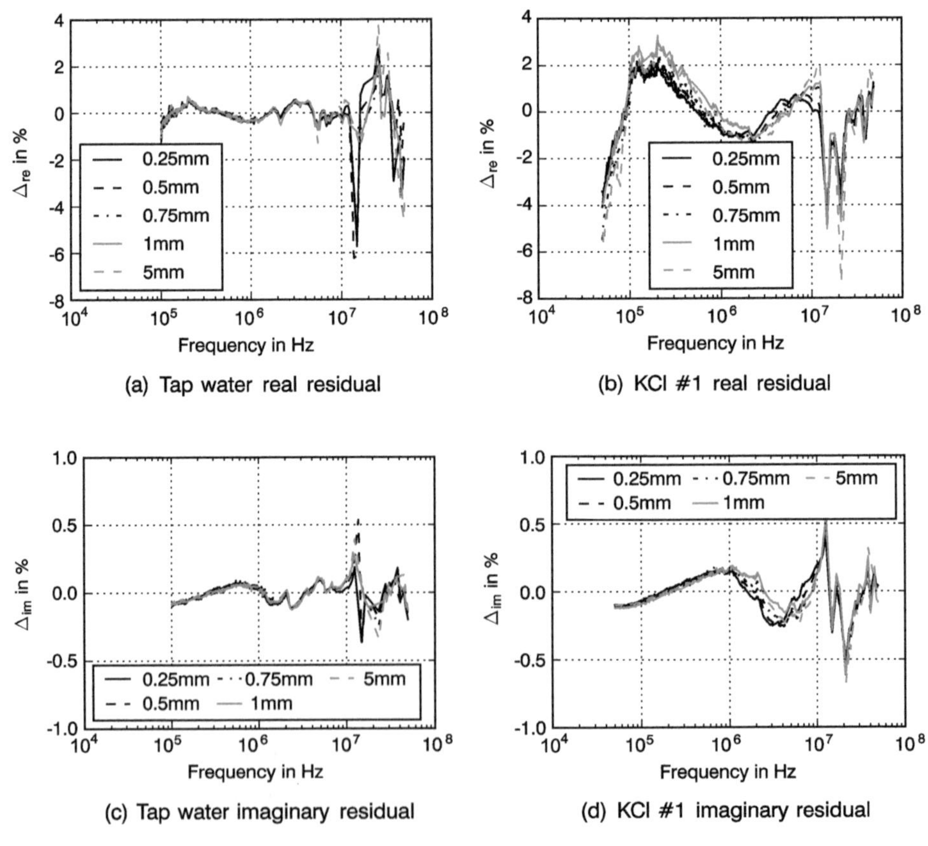

Fig. 7: Relative residuals after non-linear fitting for tap water and KCl #1.

and changes by about 10 nH, when the MUT conductivity is increased. This suggests that L_s can be solely attributed to the measurement setup itself.

The maximum residual observed for Δ_{re} is 7 %, which is a significant improvement in comparison to [5], where the maximum fitting error in the real part is up to 20 %. Significant errors arise for frequencies below 100 kHz and above 10 MHz. At lower frequencies, the measured spectrum is dominated by the CPE and small changes to its parameters cause a significant change in the modeled impedance, as can be seen in Fig. 8. Also measured signals in this region are very small, causing a decrease in the measurement accuracy, especially in $\Re(Z_{meas})$ [20].

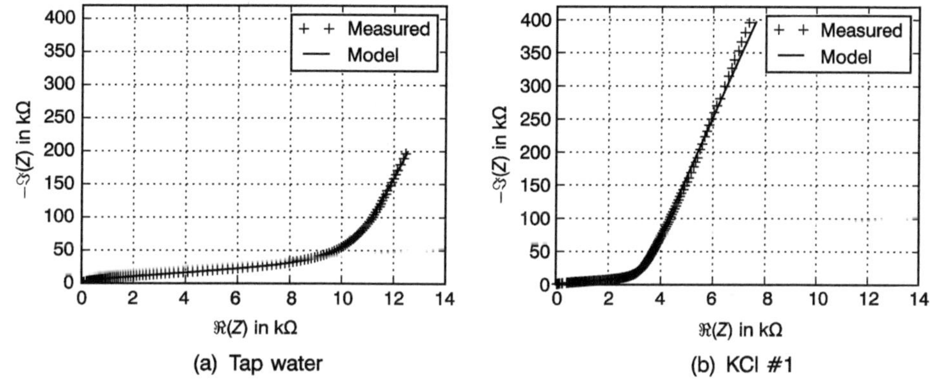

Fig. 8: Example plots of measured and fitted data for a layer thickness of 0.25 mm. For visibility reasons, only every third measurement point is shown.

Since all frequencies have an equal weight in the employed cost function, measurement errors in this region contribute equally to the fitted parameters. If frequencies below 100 kHz are excluded, Δ_{re} is below 1 % for frequencies between 100 kHz and 10 MHz, as can be seen in Fig. 7 (a). Above 10 MHz high relative residuals are observed at single measurement points. These might be caused by electromagnetic coupling into the measurement setup. Since the data smoothing is based on a linear fit, it minimizes the weighted quadratic sum of residuals, which might allow single points to pass through the smoothing scheme. To further increase accuracy, it is feasible to first extract parameters over a broad smoothed spectrum and then exclude frequency ranges with a high relative residual.

Tables 3 and 4 show the extracted values. As expected, the CPE parameters K and n do not depend on layer thickness and conductivity. Their maximum deviation is 1.72 % and 0.08 % and therefore they only represent the capacitive coupling into the MUT. Resistances decrease monotonously with increasing layer thickness and conductivity. Resistances decrease with layer thickness because of the increasing cell constant. For constant geometries the sum of R_1 and R_2 is in good approximation inversely proportional to the MUT conductivity; C_1 and C_2 increase monotonously with increasing layer

Table 3: Parameters extracted by non-linear fitting for KCl #1 for several layers.

Parameter	0.25 mm	0.5 mm	0.75 mm	1 mm	5 mm
n	−0.9928	−0.9927	−0.9926	−0.9926	−0.9927
K	$8.77 \cdot 10^{-12}$	$8.77 \cdot 10^{-12}$	$8.78 \cdot 10^{-12}$	$8.78 \cdot 10^{-12}$	$8.78 \cdot 10^{-12}$
C_1 in pF	62.0	65.1	71.1	82.9	142.0
C_2 in pF	13.0	13.8	15.0	15.3	17.1
R_1 in Ω	165.72	126.44	97.41	73.24	51.76
R_2 in Ω	2982.90	2446.66	2076.06	1980.07	1651.90
L_s in nH	251	249	249	246	251

Table 4: Parameters extracted by non-linear fitting for tap water for several layers.

Parameter	0.25 mm	0.5 mm	0.75 mm	1 mm	5 mm
n	−0.9923	−0.9922	−0.9923	−0.9922	−0.9924
K	$8.87 \cdot 10^{-12}$	$8.83 \cdot 10^{-12}$	$8.81 \cdot 10^{-12}$	$8.82 \cdot 10^{-12}$	$8.86 \cdot 10^{-12}$
C_1 in pF	77.8	77.2	80.5	81.9	167.0
C_2 in pF	12.7	13.5	14.8	15.2	17.1
R_1 in Ω	319.28	290.71	234.67	228.05	84.55
R_2 in Ω	9868.40	8318.83	6892.72	6806.33	6060.37
L_s in nH	235	240	238	236	234

thickness. This behavior can be expected because the mean permittivity increases when more water is present in the setup.

4.2 Extraction by linear fitting

In the previous section, broadband characterization has been used to extract all system parameters including the overall resistances. However, this approach fails for high conductivities because of insufficient sensitivity of the RC elements. The linear fitting procedure is independent of this sensitivity. Accuracy of the extracted CPE parameters can also be increased using the linear fitting method, because it only utilizes a measurement bandwidth, where the impedance spectrum is dominated by the CPE and the overall resistance.

Thus, for all other media the overall resistances are extracted using the linear fitting procedure. Results are shown in Table 5. Just like the resistances obtained with the non-linear fitting procedure, extracted values decrease monotonously with layer thickness and conductivity. Also, the sensitivity of R_O with respect to layer thickness decreases with increasing conductivity. This is consistent with modeling expectations formulated in [1]. For KCl #5, the sensitivity becomes small and therefore susceptible

Table 5: Extracted R_0 values.

Layer thickness	Tap water	KCl #1	KCl #2	KCl #3	KCl #4	KCl #5
0.25 mm	10465.03	3110.23	652.22	165.58	62.06	22.32
0.5 mm	8906.57	2562.32	484.37	115.95	52.42	28.17
0.75 mm	7436.25	2174.59	391.52	94.89	42.82	23.66
1 mm	7341.06	2065.30	377.92	89.09	40.56	22.32
5 mm	6442.48	1719.78	319.35	72.12	33.34	19.40

to fitting errors caused by measurement noise. This can be circumvented by averaging over several spectra.

For tap water and KCl #1 the extracted R_0 values are within reasonable tolerances equal to the sum of R_1 and R_2 obtained by the non-linear fit. The maximum error for tap water is 4.8 % while the maximum error for KCl #1 is 1.2 %. This is in line with the predictions made in Section 3.

Also the relative change of the R_0 values for a constant layer thickness is approximately proportional to the change in conductivity, i. e., if the conductivity at 0.5 mm decreases by 30 % the R_0 value increases by about 30 %. The mean deviation from this proportionality is 8.2 % while the maximum deviation is about 33 % for KCl #5 at a layer thickness of 0.25 mm. A possible cause for this is the mechanical tolerance of the measurement setup, especially at small layer thicknesses.

4.3 Comparison of the two fitting methods

Fig. 9 illustrates the difference between the two fitting procedures. The non-linear fitting method employs the entire available frequency range, the full width of the Nyquist plot, to extract the parameters of the EEC shown in Fig. 4. However, for high conductivities the non-linear extraction fails, because of the short time constants formed by low resistances.

Using linear fitting on a limited frequency range, where the Nyquist plot is a straight line, one can still obtain the CPE parameters and the overall resistances R_0 and use them to distinguish between a thin film and the actual fill level, even though the non-linear fitting method fails due to insufficient sensitivity with respect to the RC elements.

In the previous sections, it has been shown mathematically and experimentally that R_0 is related to the resistances of the EEC.

Qualitative behavior of the extracted parameters is within theoretical expectations. However, the linear fit is only valid if the frequencies used for fitting are much smaller than the cut-off frequencies formed by the conductive and capacitive part of the current. This is usually the case for highly conductive media, for which this method is intended.

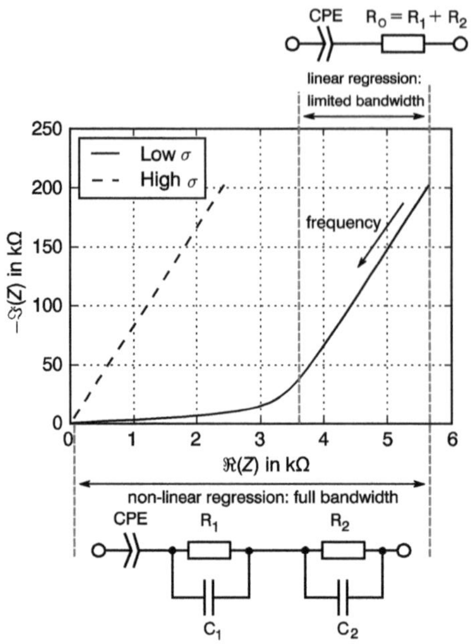

Fig. 9: Comparison of the two fitting methods. The non-linear regression utilizes the full measurement bandwidth and provides all model parameters, but fails if the bandwidth is insufficient. The linear regression utilizes only a limited portion of the measurement bandwidth but can be used even for highly conductive media, since the low-frequency information is always available.

5 Conclusion

Improved techniques for data processing have been proposed for measurement setups used to characterize thin layers of aqueous solutions. In contrast to [2], where fitting was done using purely gradient-based or direct search methods, broadband fitting of measured data to an EEC is now done using differential evolution and logarithmic compression of the parameter space.

Using differential evolution with logarithmic compression has been shown to increase reliability of conversion. The evolution algorithm can be called with larger search spaces, thus improving convenience for the user.

Also, a series inductance is introduced into the model to improve quality of fit at frequencies greater than 10 MHz. The improved method yields maximum relative errors of 0.6 % and 7 % in the imaginary and real parts of the spectrum, respectively. This is a significant improvement with respect to [5], where relative errors in the real part of the impedance were reported to be up to 20 %.

If characterization using linear fitting is sufficient, solutions with a conductivity of up to 190 mS/cm can be characterized.

Bibliography

[1] C. Weber, M. Tahedl, and O. Kanoun, "A novel method for capacitive determination of the overall resistance of an aqueous solution", In: Progress Reports on Impedance Spectroscopy, De Gruyter, 2016.

[2] T. Grosse-Puppendahl, E. M. Berlin, and M. Borazio, "Enhancing accelerometer based activity recognition with capacitive proximity sensing", In: A Nov Joi Conference on Ambient Intelligence, Springer, 2012, pp. 17–32.

[3] B. Osoinach, "Proximity capacitive sensor technology for touch sensing applications", Freescale White Paper, vol. 12, 2007.

[4] C. Weber, F. Wendler, M. Tahedl, et al., "2d-modellierung des einflusses leitfähiger schichten auf die sensorimpedanz bei kapazitiven sensoren", Tm - Technisches Messen, vol. 81, no. 9, pp. 442–449, 09/2014.

[5] C. Weber, M. Tahedl, and O. Kanoun, "Capacitive measurement setup for characterizing thin layers of aqueous solutions", In: Progress Reports on Impedance Spectroscopy, Chemnitz, Germany, De Gruyter, 2016.

[6] R. Shreiner and K. Pratt, "Standard reference materials: Primary standards and standard reference materials for electrolytic conductivity", NIST Special Publication, vol. 260, p. 142, 2004.

[7] W. Haynes, CRC Handbook of Chemistry and Physics, 93rd edition, CRC Handbook of Chemistry and Physics, Taylor & Francis, 2012.

[8] J. R. Macdonald, "Note on the parameterization of the constant-phase admittance element", Solid State Ionics, vol. 13, pp. 147–149, 1984.

[9] H. Henke, ed. "Elektromagnetische Felder", Springer, 2007.

[10] R. Storn and K. Price, "Differential evolution – a simple and efficient heuristic for global optimization over continuous spaces", Journal of Global Optimization, vol. 11, pp. 341–359, 1997.

[11] O. Kanoun, U. Tröltzsch, and H.-R. Tränkler, "Benefits of evolutionary strategy in modeling of impedance spectra", Electrochimica Acta, vol. 51, no. 8–9, pp. 1453–1461, 2006, Electrochemical Impedance Spectroscopy Selection of papers from the 6th International Symposium (EIS 2004), 16–21 May 2004, Cocoa Beach, FL, USA.

[12] T. VanderNoot and I. Abrahams, "The use of genetic algorithms in the non-linear regression of immittance data", Journal of Electroanalytical Chemistry, vol. 448, no. 1, pp. 17–23, 1998.

[13] B. L. Miller and D. E. Goldberg, "Genetic algorithms, tournament selection, and the effects of noise", Complex Systems, vol. 9, no. 3, pp. 193–212, 1995.

[14] A. S. Bondarenko and G. Ragoisha, "Potentiodynamic electrochemical impedance spectroscopy", In: A. L. Pomerantsev, ed., Electrochimica Acta , Nova Science Publ, 2005, pp. 89–102.

[15] E. Jones, T. Oliphant, P. Peterson, et al., SciPy 0.18: Open source scientific tools for Python, 2001.

[16] C. Zhu, R. Byrd, and J. Nocedal, "L-bfgs-b: Algorithm 778: L-bfgs-b, fortran routines for large scale bound constrained optimization", ACM Transactions on Mathematical Software, vol. 23, no. 4, pp. 550–560, 1997.

[17] R. Byrd, P. Lu, and J. Nocedal, "A limited memory algorithm for bound constrained optimization", SIAM Journal on Scientific and Statistical Computing, vol. 16, no. 5, pp. 1190–1208, 1995.

[18] S. van der Walt, S. C. Colbert, and G. Varoquaux, "The numpy array: A structure for efficient numerical computation", Computing in Science & Engineering, vol.. 13, no. 2, pp. 22–30, 03/2011.

[19] SymPy Development Team, Sympy: Python library for symbolic mathematics, 2016.

[20] Keysight Technologies, Keysight 4294a operation manual, Internet, Oktober 2016.

[21] B. A. Boukamp, "A linear Kronig–Kramers transform for immitance data validation", Journal of the Electrochemical Society, vol. 142, no. 6, pp. 1885–1894, 1995.

[22] M. Schönleber, D. Klotz, and E. Ivers-Tiffée, "A method for improving the robustness of linear kramers-kronig validity tests", Electrochimica Acta, vol. 131, pp. 20–27, 2014, Electrochemical Impedance Spectroscopy.
[23] U. Tröltzsch, "Modellbasierte zustandsdiagnose von gerätebatterien", PhD thesis, Universität der Bundeswehr München, 2005.

Olev Märtens, Raul Land, Mart Min, Marek Rist, Marju Ferenets, and Andres Käsper

Eddy current corrosion measurement of steel

Abstract: Contactless eddy current sensors allow efficient detection and measurement of the corrosion of metals (like carbon steel), also under the paint layer, as demonstrated in the described experiments and analysis. The corrosion was estimated by a Dodd–Deeds eddy current air-core single-coil model, according to the variation of the impedance in the sensor coil. In the case of magnetic materials (like carbon steel) the sensor signal depends not only on the conductivity of the specimen, but also on the magnetic permeability. Furthermore, the sensor lift-off influences the results, so appropriate lift-off compensation schemes are beneficial. In the current study a planar sensor coil and another ferrite-core probe were used. The estimated resolution of the corrosion layer measurement is about 1 μm (or better). A PC-based application front-end for estimation of the corroded area was developed.

Keywords: eddy current, NDT, Dodd–Deeds model, corrosion measurement, carbon steel

1 Introduction

Eddy current sensors are widely used for non-destructive testing and characterization of metals and other electrically conductive materials [1]. The analysis of the complex impedance of the single-air-core measurement coil as a probe above the metal under test has been proposed in the works of Dodd and Deeds from the 1960s [2, 3]. Alternatively, two-coil measurement setups, where one coil is transmitting the AC signal and the other is receiving, have been proposed, in one case as the receiving sensor coil is on the same side of the metal plate ("reflection-type measurement") [4] or on the other side of the metal plate under test ("through-the-metal setup") [5]. Anyway, in the current work, the single-coil configuration [2] was used, as being simple and well matching with the theory in the wide frequency range (up to 10 MHz) [6–8]. In [9], a PC-based simulation tool has been demonstrated for solving the Dodd–Deeds model for one- or two-layer metal plates with measurement coil above the metal plate. An im-

Acknowledgement: The work described in this chapter has been supported by institutional grant IUT19-11 of the Estonian Research Council and the European Regional Development Fund supporting the Estonian Center of Research Excellence EXCITE and Competence Centre ELIKO.

Olev Märtens, Raul Land, Mart Min, Thomas Johann Seebeck Department of Electronics, Tallinn University of Technology, Tallinn, Estonia, e-mail: olev.martens@ttu.ee
Marek Rist, Marju Ferenets, Andres Käsper, Eliko Competence Centre, Tallinn, Estonia

https://doi.org/10.1515/9783110558920-012

proved, computationally faster "series expansions" (instead of infinite integration in the space) version of the single-coil model is also available [10]. The Dodd–Deeds model and other similar algorithms solve the "forward problem" – that means, how the coil impedance depends on the parameters of the measured object (e. g., metal plate) and the measurement setup (parameters of the measurement coil). But in real life the solving of the inverse problem is needed – calculation of the properties (electrical conductivity and/or magnetic permeability and sensor coil lift-off) of the object under test from the measured complex impedance of the measurement coil. The lift-off effect and compensation of it have been discussed in [11]. Full inverse solutions have been proposed, e. g., in [12, 13]. One possible problem in eddy current measurements is the dependency of the results not only on the electrical conductivity of the material under test, but also on the magnetic permeability [14].

Some ideas on how to solve both these variables have been shown in [15, 16]. In [17] it has been shown (for the carbon steel AISI 1018) that the magnetic permeability is a frequency-dependent complex-valued variable. The eddy current corrosion process investigation has been done in [18, 19]. In [18] it has been shown that both the electrical conductivity and magnetic permeability of the corrosion area are much (by an order) smaller than the values for the carbon steel (for SS400 in their example).

2 Research performed

2.1 Measurement setup with air-core coil

The single-coil measurement setup is shown in Fig. 1. The setup includes the planar (PCB) measurement coil (shown in Fig. 2) with internal and external diameters D1 (2.5 mm) and D2 (16 mm), height h (20 μm), number of turns $N = 50$ and lift-off z and the metal (steel) plate with thickness of d (about 2 mm in the described experiments). The expected electrical conductivity of the material is about 6 MS/m. The impedance measurement instrument Wayne-Kerr 6500B has been used to get the coil impedance and a specially designed 3D scanner moving the probe (coil) has been developed (Fig. 3). Also the Windows-based software has been developed for reverse

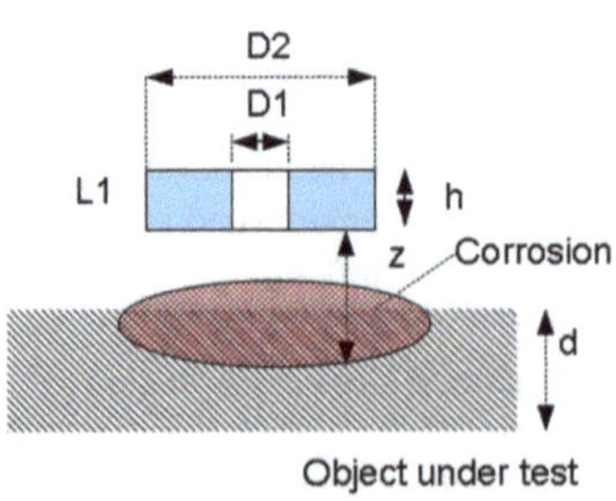

Fig. 1: Single-coil measurement setup.

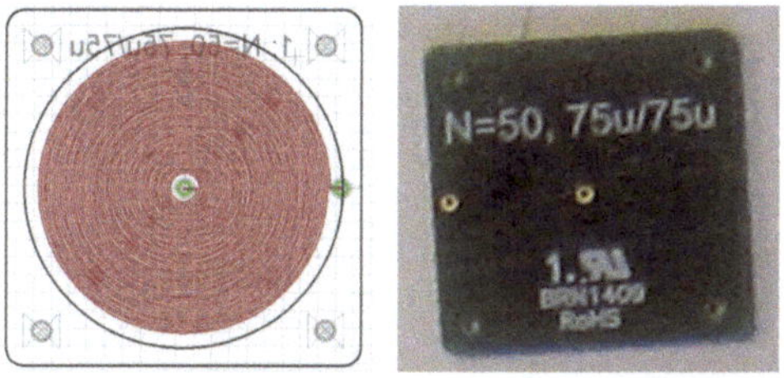

Fig. 2: Measurement coil – a photo (on the left) and PCB layout.

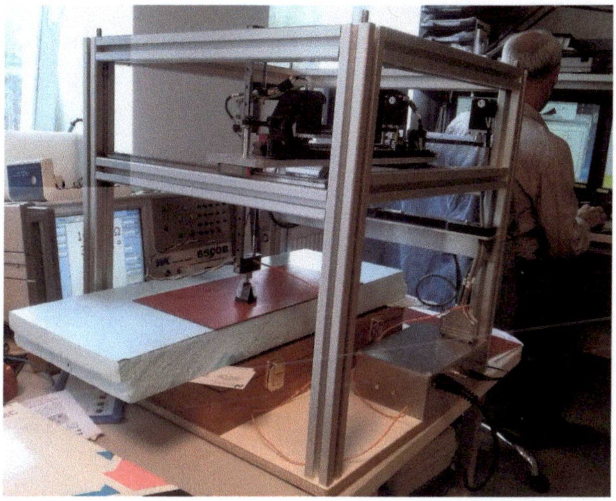

Fig. 3: Developed eddy current scanner.

eddy current modeling, using a combination of the forward model implemented as C/C++ code in the form of DLL and R-language for reverse solving, analysis and plotting of the features of the impedance measurement results. The single coil above the single-layer metal with variable lift-off and electrical conductivity and/or magnetic permeability approach has been used in this modeling and analysis software.

2.2 Experiments with initial specimen

All specimens used in the current research have been produced from the steel EN 10130 class DC01 (similar to the American AISI 1008 grade). On the first specimen (Fig. 4) some round-shape areas were corroded and the whole plate was painted afterwards. The region no. 2 was scanned by the measurement coil. The inverse-modeled results of the apparent relative magnetic permeability (μr) at 400 kHz, as given in Fig. 5, show that the corroded part is clearly distinguishable. Also, one profile line has been taken for the lift-off plots, of the same region under interest, shown as a brown vertical line in Fig. 5. Lift-off has been taken for four scans on different lift-off values, shifted vertically by 50 µm/s from each other, as shown on an upper plot of Fig. 6 (a). The same four lift-

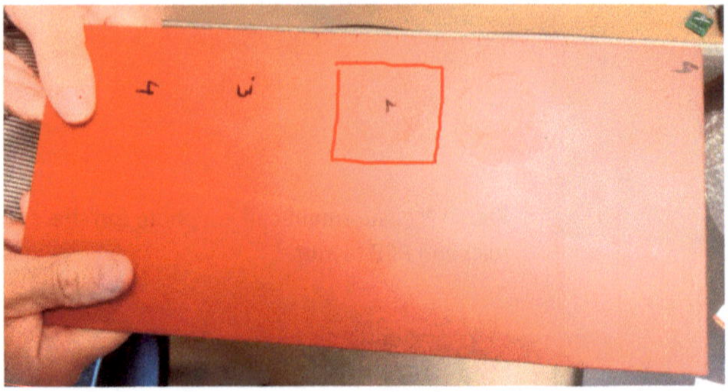

Fig. 4: The initial specimen.

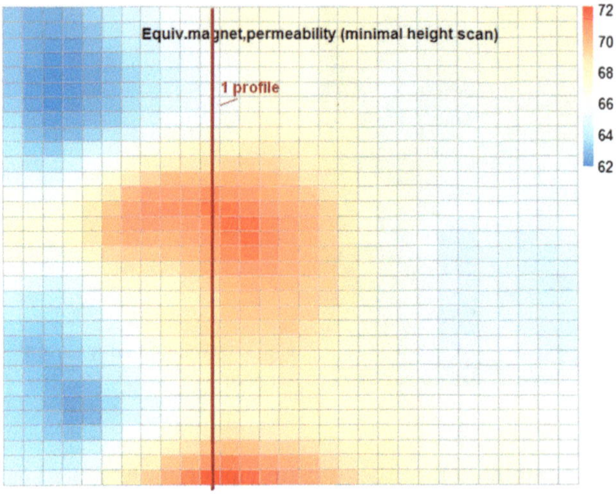

Fig. 5: Apparent relative magnetic permeability of region no. 2 at 400 kHz.

off profiles, but shifted vertically onto each other, are shown on the lower plot on the same Fig. 5. The quite exact matching of these four curves indicates the validity of the eddy current modeling and confirms the hypothesis that the corrosion depth can be characterized as change of the apparent lift-off of the material (carbon steel), in the total range of 55 µm/s for the current example.

2.3 Experiments with specimens no. 1 and 2

Two specimens have been prepared for the experimental research (no. 1 and no. 2), as shown before and after painting in Fig. 7 and Fig. 8, with round- and strip-shaped corrosion sub-areas, correspondingly. It was noticed that, if using the described air-core planar coil, the resolution of determining of the shape and area of the corroded (rust) area is limited. Still, as seen from Fig. 9, for specimen no. 2, a more or less reasonable corrosion image is acquired.

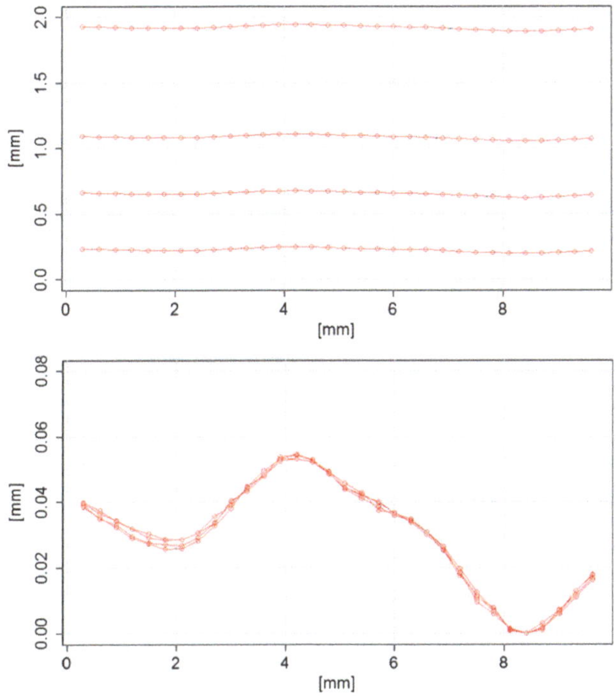

Fig. 6: Four lift-off profiles, inversely calculated from the measurements at 400 kHz; absolute lift-off values on the upper plot and relative values on the lower plot.

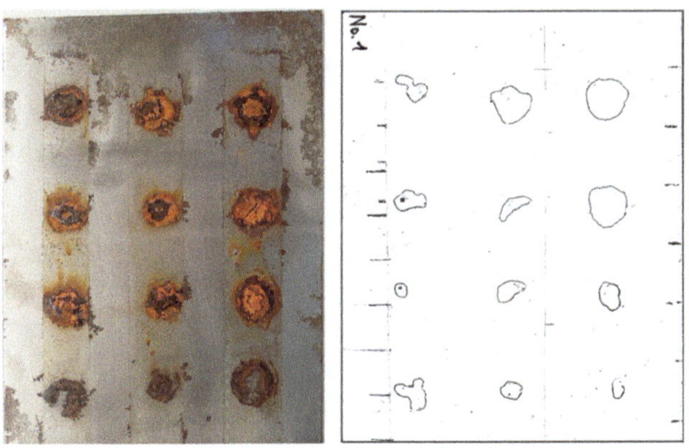

Fig. 7: Specimen no. 1, before and after painting.

3 Application to estimate the rust area

To achieve better resolution in the estimation of the rust area shape and size, a magnetic-core measurement coil (diameter of the core is 1.5 mm and $N = 50$ turns) has been introduced (Fig. 10) to work together with the mentioned 3D scanner. As, differently from the air-core approach, in real life it is hard to produce a precise eddy current model, due to several uncertainties (magnetic and geometrical properties of the coil, magnetic core, etc.) simply a complex impedance change has been taken into ac-

Fig. 8: Specimen no. 2, before and after painting.

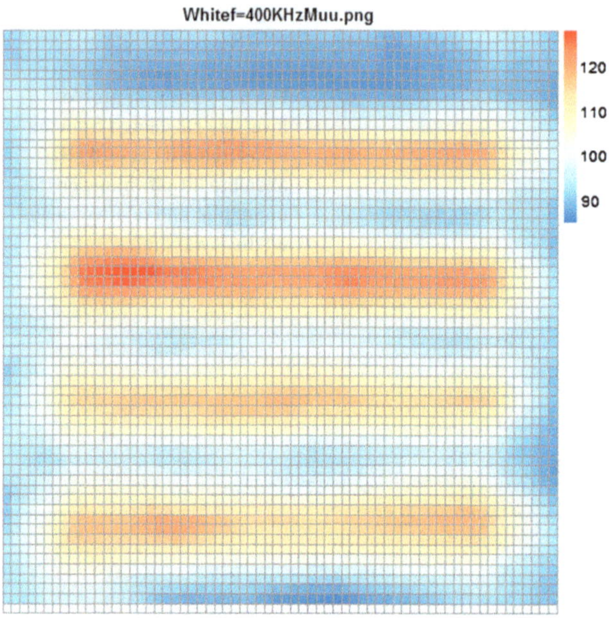

Fig. 9: Apparent magnetic permeability at 400 kHz of specimen no. 2.

Fig. 10: Sensor coil with ferrite core.

count for rust measurement. Experimentally it has been found, for the current setup and for a working frequency of 400 kHz, the linear combination of the real and imaginary components of the impedance of the coil can be used. This combined impedance values are scanned in 2D at fixed lift-off and acquired as the 2D data array; further the specially developed PC Windows-based application is segmenting this 2D plot as "corroded" and "non-corroded" parts, by selectable threshold, showing the resultant image on the screen (Fig. 11 and Fig. 12). Additionally the user can mark the rectangular region of interest on the image, where the area of the corroded segment is calculated and displayed, in absolute (mm^2) and relative (%) units.

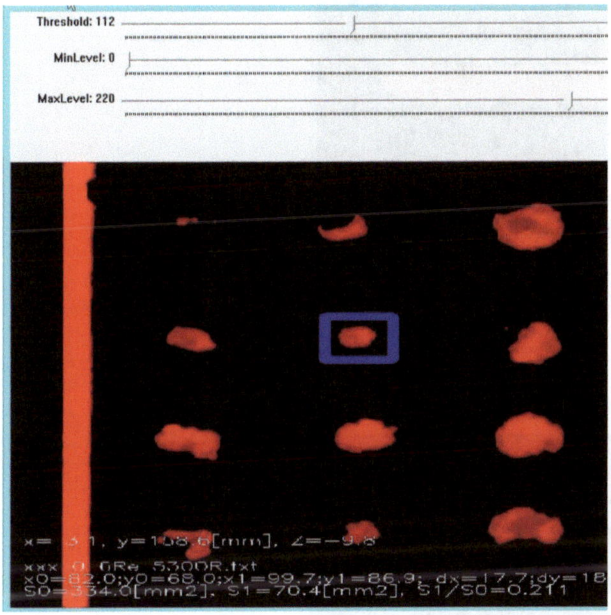

Fig. 11: Windows application for rust area measurement: specimen no. 1.

4 Conclusion and future work

The corrosion of metals, specifically the corrosion of the carbon steel, can be efficiently detected and measured by the eddy current approach with single-coil setup, also under a paint layer or similar coatings.

The proposed work is similar to [19], but instead of a pulsed sensor signal the harmonic excitation has been used, with a single, selectable frequency at a time. Also, the usage of the air-core coil in combination with the Dodd–Deeds model enables to get rid of dependency on the sensor lift-off effect and also to estimate directly, without special calibration, the depth of the corrosion.

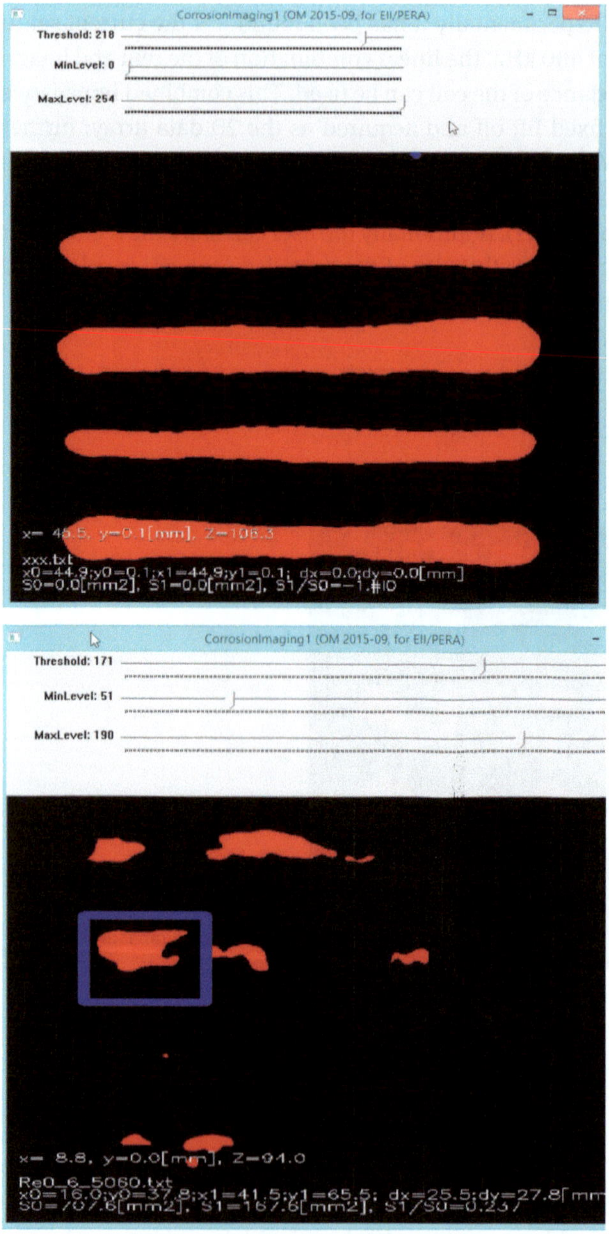

Fig. 12: Windows application for rust area measurement: specimen no. 2, at lower (left image) and higher (right image) thresholds.

The measure of the corrosion can be the variation of the apparent lift-off or change in the apparent magnetic permeability of the material under test. A resolution of 1 µm of the apparent lift-off variation can be achieved. A simple Windows application was implemented to measure the area of the corroded part(s). Further low-cost handhold instrumentation can be developed for such measurements. One future research challenge is to develop the improved absolute models for eddy current-based corrosion measurement. Also, fusion of larger planar air-core coils with smaller ferrite core probes into the same measurement procedure can combine the absolute (model-based) accuracy of the first and good spatial resolution of the second.

Bibliography

[1] J. García-Martín, J. Gómez-Gil, and E. Vázquez-Sánchez, "Non-destructive techniques based on eddy current testing", Sensors, vol. 11, no. 3, pp. 2525–2565, 2011.
[2] C. Dodd and W. Deeds, "Analytical solutions to eddy-current probe-coil problems", Journal of Applied Physics, vol. 39, no. 6, pp. 2829–2838, 1968.
[3] C. V. Dodd and W. E. Deed, "Calculation of Magnetic Fields from Time-Varying Currents in the Presence of Conductors", NASA STI/Recon Technical Report N, vol. 76, Jul. 1975.
[4] S. K. Burke and M. E. Ibrahim, "Mutual impedance of air-cored coils above a conducting plate", Journal of Physics D, Applied Physics, vol. 37, no. 13, pp. 1857–1868, 2004.
[5] C. Dodd and W. Deeds, "Absolute Eddy-Current Measurement of Electrical Conductivity", Oak Ridge National Lab., TN (USA), Tech. Rep., 1981.
[6] O. Martens, R. Land, R. Gordon, M. Min, M. Rist, and A. Pokatilov, "Precise eddy current measurements: Improving accuracy of determining of the electrical conductivity of metal plates", In: Lecture Notes on Impedance Spectroscopy: Measurement, Modeling and Applications, vol. 4, 2014, pp. 109–115.
[7] M. Parker, A. Pokatilov, T. Kubarsepp, O. Martens, and A. Kolyshkin, "Investigation of planar coil for eddy current conductivity measurements in wide frequency range", In: 19th IMEKO TC4 Symposium – Measurements of Electrical Quantities 2013 and 17th International Workshop on ADC and DAC Modelling and Testing, 2013, pp. 55–59.
[8] A. Kolyshkin, T. Kubarsepp, O. Martens, and A. Pokatilov, "Determination of electrical conductivity of metal plates using planar spiral coils", In: Proceedings of the 13th WSEAS International Conference on Systems – Held as part of the 13th WSEAS CSCC Multiconference, 2009, pp. 545–550.
[9] T. P. Theodoulidis and M. Kotouzas, "Eddy current testing simulation on a personal computer", In: 15th World Conference on Nondestructive Testing, Roma (Italy), 15–21 October 2000, Retrieved 26 May 2016, 2002. [Online]. Available: http://www.ndt.net/article/wcndt00/papers/idn228/idn228.htm.
[10] T. Theodoulidis and E. Kriezis, "Series expansions in eddy current nondestructive evaluation models", Journal of Materials Processing Technology, vol. 161, no. 1–2, pp. 343–347, 2005.
[11] M. Fan, B. Cao, P. Yang, W. Li, and G. Tian, "Elimination of liftoff effect using a model-based method for eddy current characterization of a plate", NDT and E International, vol. 74, pp. 66–71, 2015.
[12] A. Pokatilov, M. Parker, T. Kübarsepp, O. Märtens, and A. Kolyshkin, "Grid-based computational algorithm for accurate ac conductivity measurements", In: 16th International Congress of Metrology, EDP Sciences, 2013, p. 11007.
[13] S. Norton and J. Bowler, "Theory of eddy current inversion", Journal of Applied Physics, vol. 73, no. 2, pp. 501–512, 1993.
[14] J. H. Rose, E. Uzal, and J. C. Moulder, "Magnetic permeability and eddy-current measurements", In: Review of Progress in Quantitative Nondestructive Evaluation, Springer, 1995, pp. 315–322.

[15] Y. Nonaka, "A double coil method for simultaneously measuring the resistivity, permeability, and thickness of a moving metal sheet", IEEE Transactions on Instrumentation and Measurement, vol. 45, no. 2, pp. 478–482, Apr 1996.
[16] X. Ma, A. Peyton, and Y. Zhao, "Eddy current measurements of electrical conductivity and magnetic permeability of porous metals", NDT and E International, vol. 39, no. 7, pp. 562–568, 2006.
[17] N. Bowler, "Frequency-dependence of relative permeability in steel", In: AIP Conference Proceedings, vol. 820 II, 2006, pp. 1269–1276.
[18] Y. Gotoh, H. Hirano, M. Nakano, K. Fujiwara, and N. Takahashi, "Electromagnetic nondestructive testing of rust region in steel", IEEE Transactions on Magnetics, vol. 41, no. 10, pp. 3616–3618, 2005.
[19] Y. He, G. Tian, H. Zhang, M. Alamin, A. Simm, and P. Jackson, "Steel corrosion characterization using pulsed eddy current systems", IEEE Sensors Journal, vol. 12, no. 6, pp. 2113–2120, 2012.

Ingo Tobehn-Steinhäuser, Susanne Kersten,
Jaqueline Stauffenberg, Heike Wünscher, Christian Zeitnitz, and
Thomas Ortlepp

First measurements with a radiation hard relative humidity sensor for the ATLAS Experiment

Abstract: A radiation hard humidity sensor, based on an inter-digital structure is developed for the upgrade of the Large Hadron Collider (LHC) to high-luminosity LHC (HL-LHC). In this context, the interior of the detector of the ATLAS Experiment will be replaced. To guarantee a failure-free operation over the whole time, it has to be proven that the humidity in the detector region is negligible. To achieve the required very small size of the humidity sensor, double-sided wafer processing is used; furthermore the inter-digital structures consist of 3D fingers to decrease the necessary area. The otherwise customary but not radiation hard polymer as a sensitive layer is replaced by an SiO_2 matrix. In order to overcome the great distance between sensor and data processing, very compact signal preprocessing is also necessary.

Keywords: capacitive humidity-sensor, radiation hard, low residual moisture

1 Introduction

Currently the ATLAS collaboration develops a concept for the new interior detector which is completely made of silicon semi-conductor detectors [1]. ATLAS was originally an acronym for "a toroidal LHC apparatus", but is now only used as a proper name. The innermost area is formed of multiple layers of pixel detectors, further out follow strip detectors. The coolant of –40°C requires an extremely low relative humidity (RH) of the atmosphere in order to avoid condensation on the electronics in all circumstances. Therefore, a permanent monitoring of the RH and the dew point by determining moisture sensors is essential for the safe operation of the detector. (The dew point of the environment should be < –40°C). In addition to the measurement of very low amounts of residual moisture at the low temperatures, the radiation hardness of humidity sensors is a further requirement. Assuming that the ATLAS detector will operate on the high-luminosity large hadron collider (HL-LHC) for ten years, one

Ingo Tobehn-Steinhäuser, Jaqueline Stauffenberg, Heike Wünscher, Thomas Ortlepp,
CiS Forschungsinstitut für Mikrosensorik GmbH, Erfurt, Germany, e-mail: itobehn@cismst.de
Susanne Kersten, Christian Zeitnitz, Fachgruppe Physik; exp. Teilchenphysik, Wuppertal, Germany

https://doi.org/10.1515/9783110558920-013

would expect an integrated luminosity of 3000 fb^{-1}. Commercially available moisture sensors use polymers as a wet-sensitive layer and are unsuitable for use because of the lack of radiation hardness. Sensors used so far always had to be replaced after a relatively short time because they changed their characteristics [2].

2 Experimental setup

2.1 The ATLAS Experiment

For the upgrade of the LHC to the HL-LHC, the ATLAS experiment will replace the entire internal detector to handle both the high radiation load and the increased data rates. This is currently planned for the year 2022, but it is also possible that parts of the internal detector have to be exchanged in 2018. The ATLAS Collaboration is currently working on a concept for this new internal detector, which will consist entirely of silicon semi-conductor detectors. The innermost region is formed from a plurality of layers of pixel detectors, and strip detectors follow further outwards Fig. 2. Though to a different extent, both detectors have an extremely high power density of the readout electronics, which requires a correspondingly effective cooling (cooling power of 240 kW).

From Fig. 1, the requirement non-ionizing energy loss (NIEL) value is 4×10^{15} particles/cm^2. The NIEL value is a theoretical quantity, which makes it possible to easily compare different effects, which play a particular role in the damage of silicon detectors.

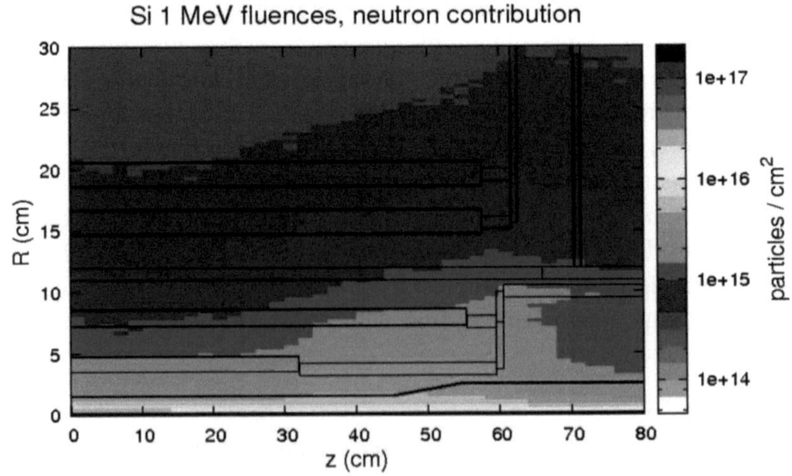

Fig. 1: The calculated non-ionizing energy loss.

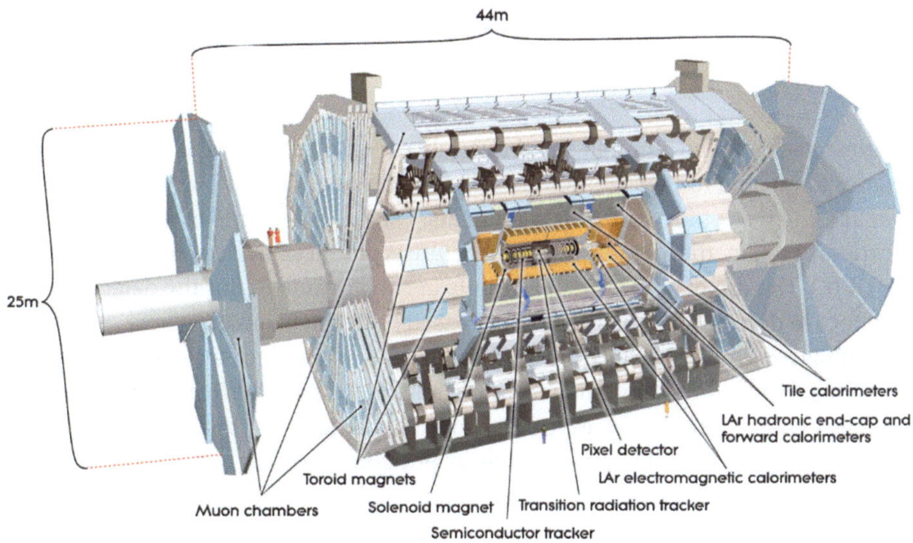

Fig. 2: The ATLAS Experiment at CERN.

An evaporative temperature of the coolant of –40°C requires an extremely low RH of the atmosphere in order to avoid condensation on the electronics in all circumstances. Therefore, a permanent monitoring of the RH or the determination of the dew point by moisture sensors is an indispensable prerequisite for the safe operation of the detector (the ambient dew point should be < –40°C). In addition to measuring the very low amount of residual moisture at low temperatures, the radiation hardness of the moisture sensors is another requirement. Based on the fact that the ATLAS detector will be operating for 10 years on the HL-LHC, an integrated luminosity of 3000 fb^{-1} is expected. The atmosphere will usually be steamed off nitrogen or highly filtered clean air without dust or oil residue, so that no contamination of the sensor by deposits is to be feared. In order to influence the radiation length as little as possible, the sensor should introduce little additional material into the detector volume or preferably consist of materials with a low mass number. For the same reason, the number of leads for supply and readout should be as low as possible. Due to the extreme packing density of the detector, the moisture sensor should also be as small as possible. Since ultimately the dew point is the critical variable, a temperature measurement should be integrated. The signal strengths must be sufficient to allow for transmission over at least 15-m cables. Depending on the type of the measured variable, local data processing is to be considered. The entire detector is located in a magnetic field of 2 T.

The whole arrangement is designed for a duration of ten years. After the installation, there is no access capability, which consequently places the highest demands on the reliability of all installed components. Maximum stability of all calibration factors is necessary, unless the sensor system has a self-calibration.

2.2 Design of the radiation hard humidity sensor

2.2.1 Layout

In order to achieve the required small size, silicon-on-insulator (SOI) material is used in order to be able to perform double-sided processing. This allows to use both the front and the rear of the chip for components. On the device side of the wafer with a thickness of 10 μm, the 3D inter-digital structure is produced by deep reactive ion etching (DRIE). The use of a 3D structure results in two advantages for the detector. First, the effective detector area is thereby increased with the same base area. Second, it is advantageous to use highly doped silicon instead of metal for the conductive structures. This is in accordance with the requirement to use materials with a small number of nuclear charges. This reduces the activation of the material by the irradiation and the radiation length is influenced as little as possible by the sensor.

Three different variants have been implemented with regard to the geometric layout. The combinations listed in Fig. 3 are used for the finger widths and distances between the fingers.

gap width [μm]	gap depth [μm]	finger width [μm]
1.1	10	3.5
1.5	10	2.5
3.0	10	4.0

The smaller the distance between the fingers, the higher the total capacitance of the sensor. On the other hand, small finger distances make it difficult to insert the moisture-sensitive material between the fingers. Since this is an RH sensor, the temperature must be measured simultaneously to determine the dew point temperature. This is done according to the formula

$$T_T = f_R^{\frac{1}{8.02}}(109.8 + T) - 109.8, \tag{1}$$

where T_T is the dew temperature point and f_R is the RH.

Like all other components, the diode, used to measure the temperature, is placed on the handle side of the wafer. In the first design, a micro-controller is used, which has very small geometric dimensions [3]. This concept is followed because the sensor must first be as small as possible, but on the other hand the output signals should be given via a 16-m long line (Fig. 4). In the case of the inter-digital structure, a square pulse is applied to the structure via a fixed resistor. On the basis of the capacitance, which has been adjusted due to the humidity, there is a decay time due to the RC constant.

This is analyzed by measuring the voltage after a fixed time t. In order to provide the information of this analog signal via the signal line, it is digitized with an ADC integrated on the micro-controller. The micro-controller used is commercial standardware

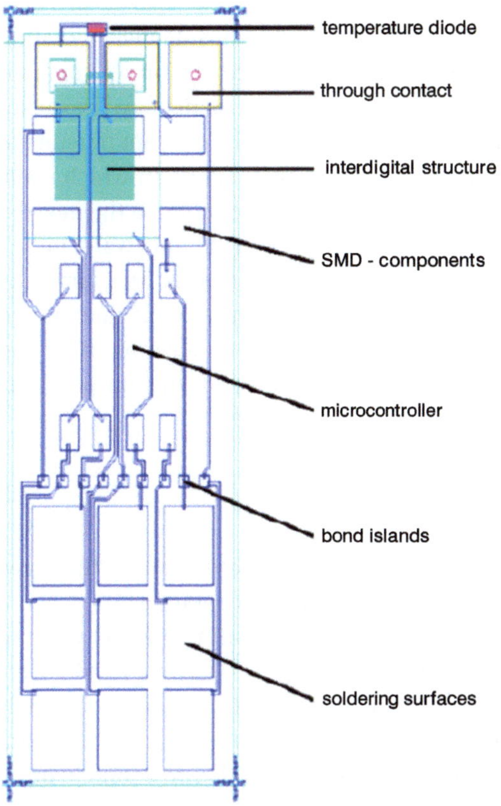

Fig. 3: Layout of the sensor.

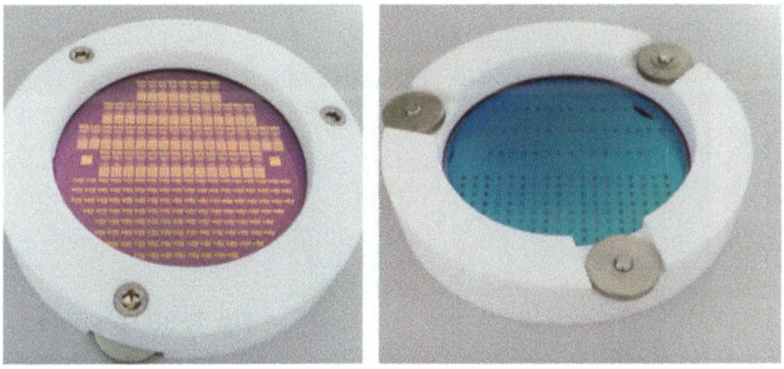

Fig. 4: Double-sided preparation of the wafer.

and therefore does not meet the increased requirements at CERN, in particular the very high radiation load. Therefore, this part is only a conceptual design and should be transferred from a sub-center of CERN to a layout with radiation-proof electronics. The large solder pads in the layout are only for programming the micro-controller and do not belong to the definitive concept of the sensor. For the online calibration of the system, the installation option of a capacitor is also provided as a reference on the circuit board.

All previous RH sensors used in CERN failed due to their insufficient radiation resistance. The moisture-sensitive layer consists of polymer chains on all commercially available sensors. These are split at high beam exposures. In this way, the sensor changes its response behavior and thus its characteristics. In the concept pursued at CiS, SOL-GELs are used. After processing, these form an SiO_2 matrix in which some water binding substances are additionally incorporated. Furthermore, in the new development of the SOL-GEL used, no organic solvents were used but only water [4].

In dip coating, a sample is placed in a liquid medium to coat it. The sample is dipped into the substance, then remains there for a certain time t and is then pulled out at a constant velocity v_0. After this, the sample has to dry for a certain time and may have to be dried again in an oven at higher temperatures. For this purpose, the layer thickness can be calculated using the Landau–Levich, i. e.,

$$d = 0.8\sqrt{\frac{\eta v_0}{\rho g}}. \tag{2}$$

The layer thickness d is determined by this formula from the viscosity η, the drawing speed v_0, the density ρ of the liquid and the gravitational constant g [5].

For applying the SOL-GEL to the sensor layer, the spin coating method was unsuitable since the 3D sensor structures were not filled (Fig. 5). The best results were obtained by the dip coating method, whereby the structure to be coated is slowly drawn out of the solution and subsequently annealed. It was also found to be advantageous during the immersion phase to sonicate the solution with ultrasound in order to achieve a true homogeneous coating.

Fig. 5: 3D structure.

2.3 Measurements

2.3.1 Uncoated sensors

The sensors were initially measured dry and with de-ionized water using an LCR meter (Agilent 4294A). The results are shown in Fig. 6.

Significant differences of the sensor signal at room temperature in the dry state and with the applied drop of de-ionized water can be seen. The water quantity

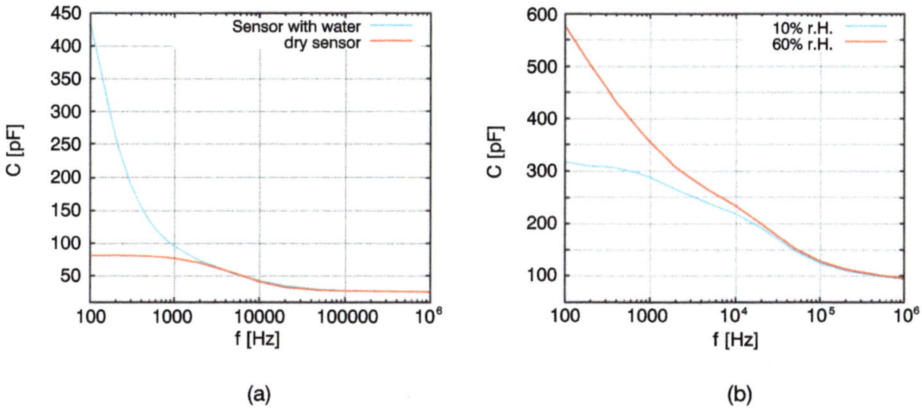

Fig. 6: (a) Uncoated Sensor. (b) Coated Sensor.

was 5 µL. The difference corresponds to about 350 pF at 100 Hz and is thus easily detectable. In this way, the general functionality of the first built-in sensors was checked. After this function test, the sensors were provided with the moisture-sensitive layer for further measurements.

2.3.2 Coating with dip coating

The temperature was kept constant at 25°C during the measurements and the RH varied between 10 % and 60 %. First, the capacitance was measured as a function of the frequency. The measured frequency range was between 100 Hz and 1 MHz. Fig. 7 shows the difference between the two moisture values.

Based on these measurements, a value of 5 pF/ % RH at 100 Hz can be calculated. The capacity at 100 Hz and 10 % RH is 317 pF and with a humidity of 60 % 575 pF is

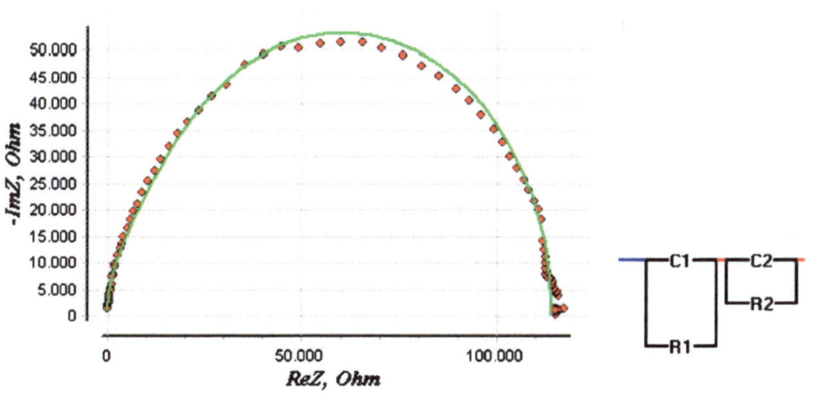

Fig. 7: Equivalent circuit.

achieved. In addition, the impedance and the phase were measured in a frequency range of 25 Hz to 1 MHz. Again in this case, the difference between the different moisture values can be clearly seen. At 10 % RH, the impedance is 6.89 MOhm and with a humidity of 60 % 3.71 MOhm. The course of the measurement curves corresponds to that of RC elements.

2.3.3 Irradiation of samples

To investigate whether the new layer used will withstand future radiation exposure without changing its required properties, several samples were irradiated at the 27-MeV proton linear accelerator in Birmingham [6]. Prior to irradiation and after irradiation, FTIR spectra were recorded to investigate whether the chemical bonding structures detectably altered. In order to obtain a better indication of the stability of the coating, different samples were irradiated at different doses. These had the values 0.77 MGy, 1.11 MGy, 4.84 MGy and 6.12 MGy. Even at the highest dose, no significant differences were found between irradiated and unirradiated specimens. In a next step, a mounted unit under irradiation is to be subjected to a functional test.

3 Summary

The experimental results so far suggest that the concept underlying the sensor works reasonably. The vias proved to be critical; the technology still requires a modification to make it less susceptible. Moreover, it was found that the method of dip coating for applying the sensitive layer appears to be the only practicable method. For this purpose, the process parameters such as pulling speed and tempering times had to be found by means of corresponding test series. Further, an additional application of ultrasound during the coating process was useful to achieve a more homogeneous coating image. Experiments with spin coating showed unsatisfactory results because the 3D structures are not filled sufficiently.

Bibliography

[1] CERN Document server, https://home.cern/about/experiments/atlas.
[2] Honeywell, HIH-4000 Series, https://sensing.honeywell.com.
[3] Flash-Based 8-Bit CMOS Microcontrollers PIC12F617 8-Pin, http://www.microchip.com/.
[4] N. Winkler, “GmBU Gesellschaft zur Förderung von Medizin”, Bio- und Umwelttechnologien.
[5] I. Buchem, “Nanostrukturierte Aluminiumfluoridschichten, Über das neuartige Niedertemperatur Sol-Gel Verfahren und die charakteristischen Eigenschaften”, Dissertation Verlag; Dezember 2010.
[6] AIDA2020, “Advanced European Infrastructures for Detectors at Accelerators”.

Frank Wendler, Paul Aurosmit, and Olfa Kanoun

CNT-enhanced carbon electrode for nitrite detection and water analysis

Abstract: In this contribution we present a novel electrode based on a bare pencil lead electrodes (PLEs) coated by carbon nanotubes. We investigate the sensitivity of the novel electrode during electrochemical analysis of aqueous solutions. For this purlose, a phosphate buffer having different nitrite concentrations has been subjected to meaurements with enhanced and bare PLEs over a wide potential range. The proposed coating increases the sensitivity to nitrite by increasing the electrode surface and its reactivity and shows therefore a significantly observable and sensitive signal.

Keywords: square wave voltammetry, nitrite, pencil electrode, carbon nanotubes

1 Introduction

Nitrite ions (NO_2^-) can be found in food, water and soils [1]. In the food industry, nitrite is used in the curing of meat by preventing bacterial growth [2]. It is found as a natural component in vegetables and water due to chloramination [3]. High concentrations of nitrite ions are toxic to humans and animals. Especially during cooking, excess nitrite in meat can react with degradation products of amino acids to form carcinogenic nitrosamines [1, 4–6]. It also oxidizes hemoglobin to methemoglobin, which causes health problems like methemoglobinemia [7].

To detect harmful concentrations of nitrite in the food chain of human and animals, a reliable and cost-effective way of determining the concentration of nitrite in an aqueous solution is needed. The method of square wave voltammetry (SWV) provides excellent results in the quantitative analysis of chemical compounds in low concentrations [8]. The selective nature of this method is very useful to detect chemicals that are involved in electrochemical redox reactions. The sensitivity and the selectivity of the method can be further enhanced by specialized electrodes.

This enhancement is usually performed by coating a bulk electrode material within an electrocatalytic layer. Often used bulk materials for the electrodes are glassy carbon and pyrolytic carbon. These are expensive and are usually cleaned and re-used in several experiments. To evaluate a large amount of coatings in the optimization process a more cost-effective base electrode is needed.

In this work pencil lead electrodes (PLEs) are proposed as cheap and accessible base material for electrode prototyping. In an additional stage the electrodes are

Frank Wendler, Paul Aurosmit, Olfa Kanoun, Chair for Measurement and Sensor Technology, Chemnitz University of Technology, Reichenhainer Strasse 70, 09126 Chemnitz, Germany, e-mail: frank.wendler@etit.tu-chemnitz.de

https://doi.org/10.1515/9783110558920-014

coated with multiwalled nanotubes (MWNTs) to enhance the sensitivity to the concentration of nitrite ions as they have a strong electrocatalytic property [9]. The aim of the paper is to invest the feasibility of this type of electrodes and to evaluate its sensitivity for measurements.

2 Experimental

2.1 Solution preparation

For the determination and graphical representation of the reaction kinetics, SWV was used.

SWV is known for well-defined current peaks at characteristic redox potentials [8]. It is also very sensitive to the electrode kinetics [8]. The redox reaction that takes place at the electrode is

$$NO_2^- \longleftrightarrow NO_2 + e^- . \tag{1}$$

A 0.65-M (pH = 7) phosphate buffer solution (NaH_2PO_4/Na_2HPO_4) from Grüssing GmbH and sodium nitrite of 99 % purity was used to prepare the investigated solutions. The buffer solution was diluted to 0.325 M; 100 mL of the diluted buffer solution was contaminated with 0.7 mg sodium nitrite and 100 mL of the buffer solution was left uncontaminated. Both base solutions were used to create samples with mixing ratios of 0 %, 25 %, 50 %, 75 % and 100 %. In this 100 mL of base solution the nitrite concentration was 100 µM. So in this investigation the different samples of 100 %, 75 %, 50 %, 25 % and 0 % correspond to concentrations of 100 µM, 75 µM, 50 µM, 25 µM and 0 µM.

2.2 Preparation of electrodes

The pencil lead consists of graphite, clay and wax has a resistance of approximately 6 Ohms. The whole pencil was heated to reduce the resistance and evaporate the excess binding material so that the properties of graphite are predominant. The pencil lead is covered by an insulating cover so that only the tip of the pencil is exposed to the solution.

For the enhancement, multiwalled carbon nanotubes (MWCNTs) of 2 % w/w in 2-propanol (stabilized) were used from Future Carbon GmbH. The pencil lead was cleaned and then the tip was dipped in the prefabricated MWCNT solution.

The pencil lead was dried in a heater for 5 hours to evaporate the propanol and fixate the MWCNTs on the surface of the pencil lead.

2.3 Instrumentation and experimental setup

A three-electrode cell was used for the electrochemical experiments. The measurements were conducted using the HEKA 310 potentiostat. The pencil lead was used as the working electrode. Ag/AgCl 0.1 M KCl from CH Instruments, Inc. was used as the reference electrode. The counter electrode consisted of the platinum wire. The measurement setup is described in Fig. 1.

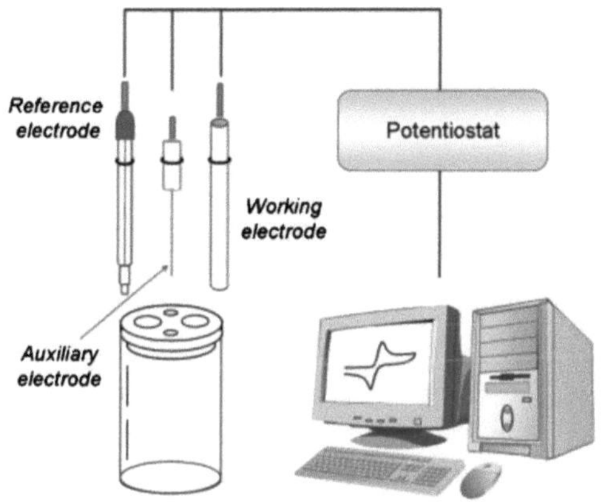

Fig. 1: Experimental setup [10].

2.4 Instrument settings

The start potential was set at −0.5 V and the end potential at 1.2 V. During the measurements in the negative potential range the samples were able to reach a steady state. The potential was increased in steps by 10 mV, and the square wave frequency was set to 1 Hz. The scan rate was set at 10 mV/s. The current was sampled from 98 %–100 % of the period so that the double layer capacitive effects can be avoided and only the Faradaic currents are represented in the data.

3 Results and discussion

The electrode was immersed in all five samples solutions and the SWV was conducted. The bare PLE in Fig. 2 shows a small current peak in the region of 850 mV. This peak is only visible at high concentrations of 100 % and 75 % of the base nitrite solution. We conclude that the bare PLE has a limited use in the detection of nitrite, but can be used as base electrode to observe the effects of enhancing coatings.

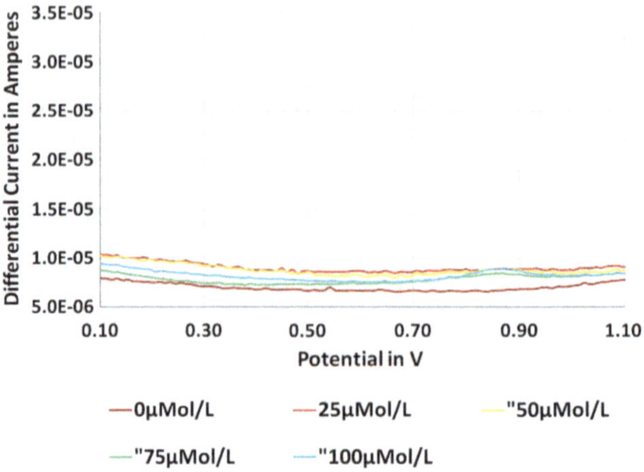

Fig. 2: SWV of pencil lead at different concentrations of nitrite.

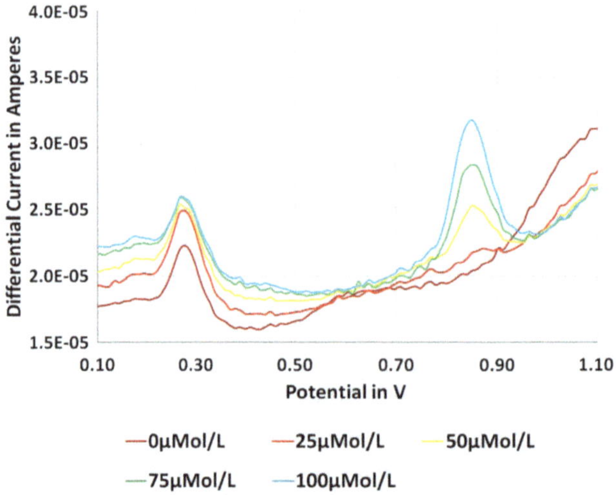

Fig. 3: SVW of CNT-coated PLE at different concentrations of nitrite.

The PLE was coated with the MWNT film and immersed in the same five sample solutions. The coated electrodes show an increased current at all concentrations as seen in Fig. 3 due to the increase in surface area of the nanomaterial coating. The current peak at 0.85 V for the nitrite is enhanced and also observable at lower concentrations of nitrite, as shown in Fig. 4. This increases the sensitivity and the detection limit for nitrites of this electrode type. Furthermore, a second peak can be observed at 0.275 V. This peak has a constant level at all concentrations of nitrite. All the five samples always contain the same amount of phosphate buffer and de-ionized water irrespective of the nitrite concentration. So we conclude the peak at 0.275 V is related to the con-

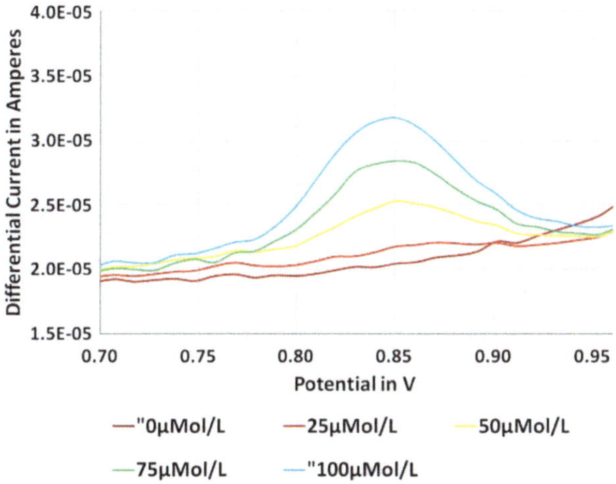

Fig. 4: SWV of CNT-PLE showing the peak currents at different concentrations of nitrite.

centration of the phosphate buffer. Since the SWV waveform for the bare PLE does not show any prominent peak at any potential, we assume that CNTs have a catalytic effect on the phosphate in the buffer solution.

The direct comparison between the voltammograms of the bare PLE and CNT-enhanced PLE (CNT-PLE) for base solutions with concentrations of 0 % and 100 % of nitrite can be seen in Fig. 5 and Fig. 6, respectively. The peak related to the base solu-

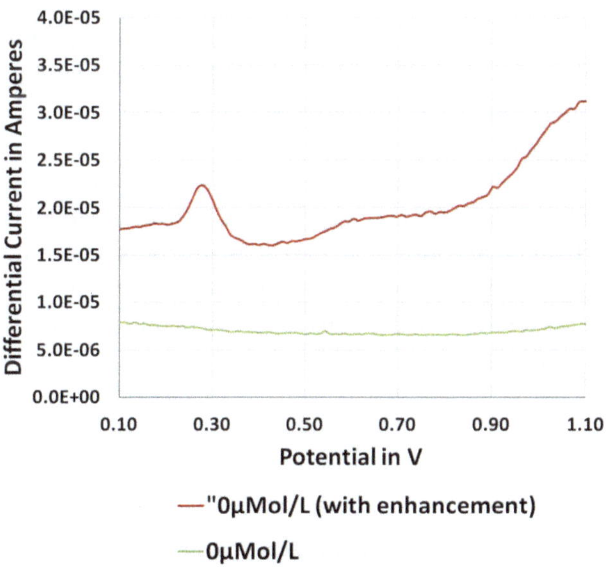

Fig. 5: SWV of PLE and CNT-PLE in 0 % nitrite solution.

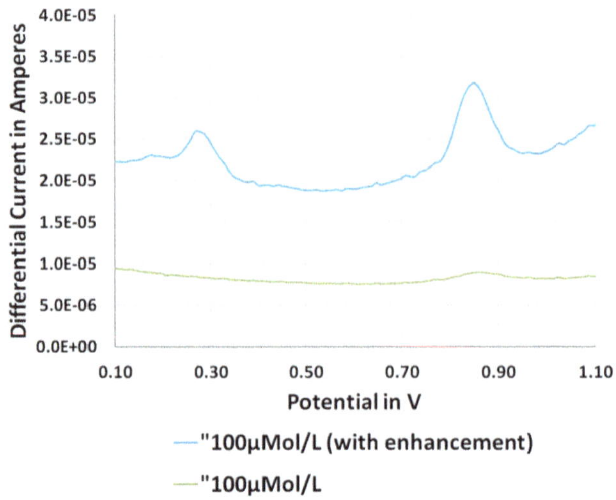

Fig. 6: SWV of PLE and CNT-PLE in 100 % nitrite solution.

tion at 0.25 V is observable in both voltamogramms for the enhanced electrode. The nitrite peak at 0.85 V can be observed with both electrodes in solutions containing nitrite.

4 Conclusion

In this chapter, a novel CNT-coated PLE is proposed for nitrite measurements by SWV. The results show that the PLE can be used as an excellent base electrode for modification and evaluation. The modification by coating with CNTs show good results for concentrations of nitrite from 100 μM to 25 μM.

The CNT-coated electrode can be made sensitive to other chemicals, like phosphate, to adapt to other chemicals and fields of application. The electrode material is cheap and can be enhanced with minimal effort. Therefore PLEs support a screening of large quantities of coating materials. For future work the miniaturization and optimization will be the main aim along with checking the sensitivity to other reagents, so that cheap and reliable electrochemical sensors can be created.

Bibliography

[1] Y. Wang, E. Laborda, and R. G. Compton, "Electrochemical oxidation of nitrite: Kinetic, mechanistic and analytical study by square wave voltammetry", Journal of Electroanalytical Chemistry, vol. 670, 2012.

[2] E. F. Binkerd and O. E. Kolari, Food and Cosmetics Toxicology, vol. 13, no. 6, pp. 655–661, 1975.
[3] N. G. Hord, Y. Tang, and N. S. Bryan, The American Journal of Clinical Nutrition, vol. 90, no. 1, pp. 1–10, 2009.
[4] B. P. Kozub, N. V. Rees, and R. G. Compton, "Electrochemical determination of nitrite at a bare glassy carbon electrode; why chemically modify electrodes?", Sensors and Actuators B, vol. 143, 2010.
[5] C. Yanping, Y. Changzhu, W. Zeng, M. Oyama, and P. U. Wenhong, "Electrochemical determination of nitrite using a gold nanoparticles-modified glassy carbon electrode prepared by the seed-mediated growth technique", Journal Analytical Sciences, vol. 23, 2007.
[6] P. Manas and G. Vellaichamy, The Analyst, vol. 135, pp. 2711–2716, 2020.
[7] L. Fu, Y. Shuhong, L. Thompson, and Y. Aimin, RSC Advances, vol. 5, pp. 40111–40116, 2015.
[8] F. M. Cubius, A. Ispas, A. Bund, and P. Ilea, "Square wave voltammetric detection of electroactive products resulting from electrochemical nitrate reduction in alkaline media", Journal of Electroanalytical Chemistry, vol. 675, 2012.
[9] S. Timur, U. Anik, D. Odaci, and Lo Gorton, "Development of a microbial biosensor based on carbon nanotube (CNT) modified electrodes", Electrochemistry Communications, vol. 9, no. 7, July 2007.
[10] A. Alvarez-Lueje, M. Perez, and C. Zapata, Pharmacology, Toxicology and Pharmaceutical Science, 1st edition, 978- 953-51-0099-7, 2012.

Part IV: **Bioimpedance measurements and methods**

Rauno Gordon, Mart Min, and Raul Land

Energy-efficient pulses for deep brain stimulation

Abstract: Energy-efficient excitation of neural tissue is investigated in this paper through the generation of optimal shape pulses. It is essential for neural stimulation to have a large percentage of the energy in the most useful frequency range. The useful range is above the frequency of the tissue–electrode interface capacitance and below the frequency of other parasitic capacitances (small capacitances of cables, connectors). One powerful observation to determine that useful range is that it centers on the minimum phase angle of the impedance. In this study we have generated stimulation signals in the form of short chirp pulses that have 95 % of energy in one decade of frequency range.

1 Introduction

Electrical stimulation of neural tissue (deep brain stimulation) is an established method for the treatment of neural diseases. The American Food and Drug Administration (FDA) approved deep brain stimulation as a treatment for essential tremor in 1997, for Parkinson's disease in 2002 and for dystonia in 2003 [1]. For treatment, stimulation leads are placed in the brain according to the type of symptoms to be addressed.

Practical application of the method needs extremely high energy efficiency. The power of the pulses has to be exactly controlled because the tiny electrodes with high electrical impedance can degrade themselves but they can also heat up the nerve tissue, damage it and modify the neural functioning. Nowadays mostly fully implanted devices are used for neural tissue stimulation (battery-powered heart and brain pacemakers). Their functions include continuous monitoring, signal processing, decision making and stimulation when required. They should be able to operate for years. Therefore simple and extremely low-power electronics is required for the generation of stimulation pulses with optimized waveforms.

The spectra of the stimulation pulses need to be matched with the spectral properties of the complex electrode–tissue impedance $Z(j\omega)$ for the desired neural tissue stimulation effect. The energy of the stimulation signal should desirably be concentrated into a spectrum band between indicative frequencies f_1 and f_2. We should avoid generating energy with frequencies under the lower bound f_1 where the stimulation signal can poorly penetrate into the tissue due to double-layer impedance on the electrode interfaces. The frequency above f_2 should again be avoided because there are

Rauno Gordon, Mart Min, Raul Land, Thomas Johann Seebeck Department of Electronics, Tallinn University of Technology, Tallinn, Estonia, e-mail: rauno.gordon@ttu.ee

https://doi.org/10.1515/9783110558920-015

parasitic losses due to small capacitances and inductances in the whole stimulation pathway.

2 Objectives

Sequences of pulses with the repetition rates from 100 Hz to 200 Hz and durations between 100 μs to 500 μs are typically used for stimulation [2]. Our aim is to find specific waveforms for the stimulating pulses that have most of the energy in the most sensitive frequency range for the neural tissue. We try to focus as much energy as possible to a narrow energy spectrum band. The center of the frequency band would be the frequency where the impedance has its maximum imaginary part – also called the peak resistance frequency f_p [3, 4]. Our metric for measuring the focussed energy is the percentage of the energy that falls into one decade of the spectrum (Fig. 1). For the most promising excitation signal we also look at the spectrum of repeating pulses with various repetition rates to see if a practical usage scenario can affect the spectrum and therefore the energy efficiency of the stimulation.

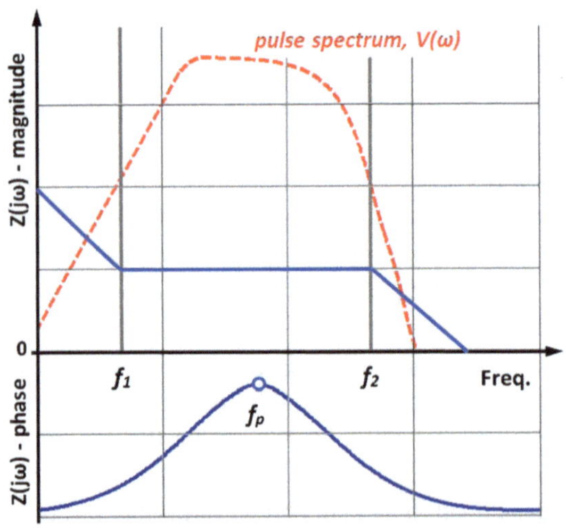

Fig. 1: Simplified magnitude and phase of tissue impedance (blue solid lines) with desired pulse spectrum (red dashed line). Frequencies f_1 and f_2 are bounds between which we want to focus the energy of the pulse; f_p is the frequency where impedance has its minimum phase angle (peak).

3 Methods

Excitation pulses for nerve tissue must be exactly balanced – for a single excitation pulse there should be a positive and a corresponding negative part of the signal where electric charge given into the electrode is later balanced. If unbalanced charge is left into the tissue near the electrode, tissue damage can occur over time and electrode degradation can follow too, which would limit the longer-term useability of the device [5].

We have previously studied square pulses (also balanced) with and without a gap between the excitation and the balancing pulse [6]. Square pulses have the advantage of being easy to generate with simple low-power electronics. We looked into square pulses with various ratios between the amplitude and longevity of the excitation pulse and balancing pulse. We found that square pulses can be generated with up to 85 % of energy in the desired band within one decade. A good and typical example is given in Fig. 2. A gap is often added in practice between the excitation and balancing part of the pulse [7]. Based on our results it did not improve the energy efficiency of the pulse but degraded it, depending on the length of the gap [6].

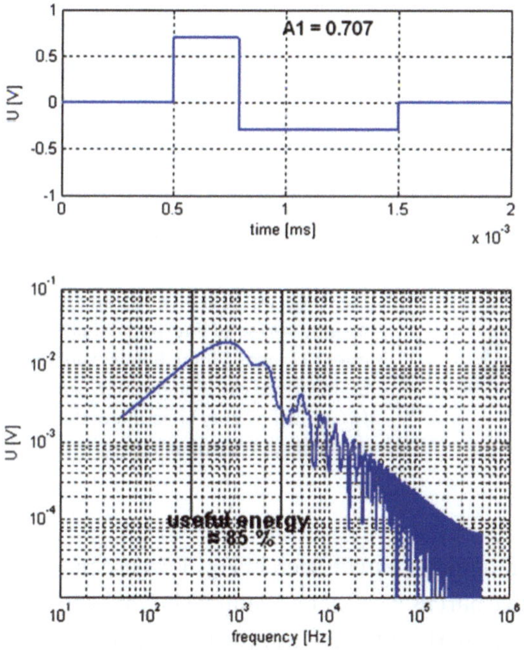

Fig. 2: Above: balanced square excitation pulse with an amplitude ratio of 0.707. Below: spectrum of the signal calculated with discrete FFT showing the bounds with gray lines where useful signal energy is calculated.

The stimulating pulses with the predetermined repetition rate and duration are synthesized mathematically for obtaining the focused and controllable energy spectrum. Several pulse signals with a frequency that rises following square function are experimented with.

Pulses of the first type are produced with frequency rising following a square function and when the sinus signal reaches the quarter-period, the frequency is lowered following the same square function. This however produces a signal that is not symmetrical with respect to the central line where the frequency rise was reversed. The

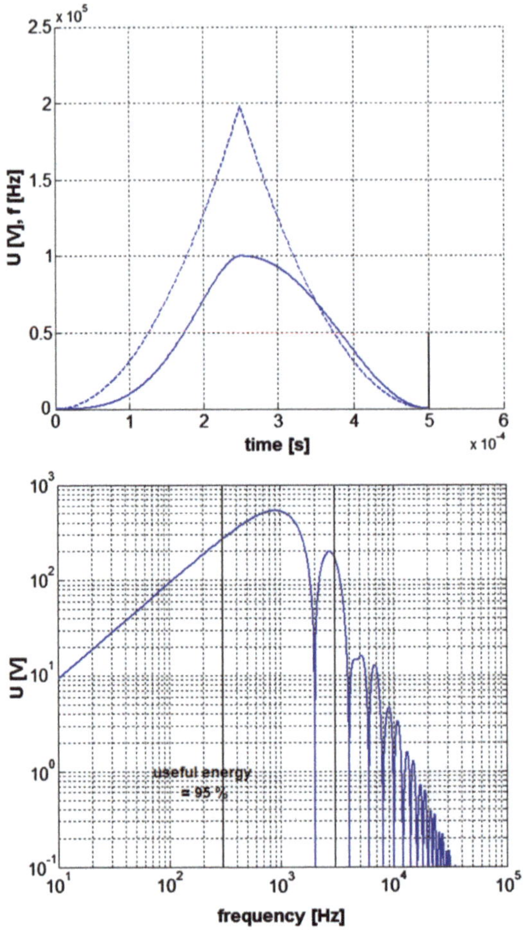

Fig. 3: Above: part of balanced signal magnitude (solid line) and rising frequency (dashed line). Below: spectrum of the signal calculated with discrete FFT shown by gray lines where the signal energy is calculated.

frequency function and an example of the pulse can be seen in Fig. 3. Please note that the pulse includes a balancing, negative part as well that is identical but not shown in the figure. After generating the balanced parts of the pulses, the frequency spectrum was calculated with discrete FFT in MATLAB. Energy within one decade of frequency was calculated for comparison purposes. The selection of the actual desired frequency band depends on the application but here the one-decade band is chosen based on typical tissue conditions.

Pulses of the second type use the rising frequency as well, but the half-period of the signal is used. The same signal in reverse sequence is then added for the balance on the negative side. The rising frequency function and the best sample pulse with both positive and negative parts is shown in Fig. 4. The spectrum and energy in one decade were calculated after that.

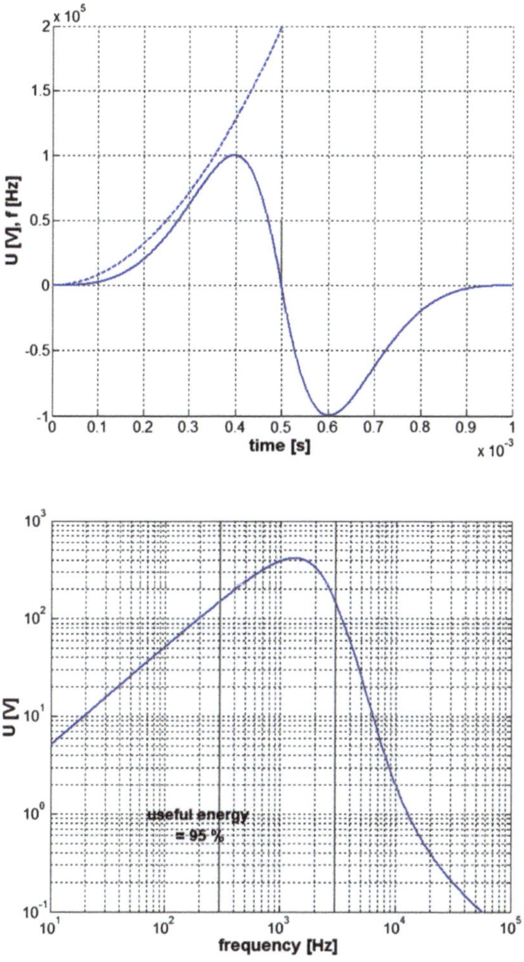

Fig. 4: Above: balanced signal magnitude (solid line) and rising frequency (dashed line). Below: spectrum of the signal calculated with discrete FFT shown by gray lines where the signal energy is calculated.

4 Results

The spectrum and energy efficiency of the first type of balanced signal are given in Fig. 3. We selected one decade in frequency to be the band where we desire to concentrate the energy. It can be seen that 95 % of the energy resides in the desired region.

The spectrum and energy efficiency of the second type of the pulse are given in Fig. 4. The frequency spectrum exhibits a smooth and narrow curve that also concentrates around 95 % of energy into the desired band.

The second type of pulse – half-cycle chirp pulse with added identical but reversed balancing part – shows a remarkable and smooth drop in energy at higher frequencies (Fig. 4). This makes it especially suitable for stimulation when parasitic losses at higher frequencies are limiting. We also investigated the spectrum of this second type of pulse in situations where the stimulation needs to be repeated. The

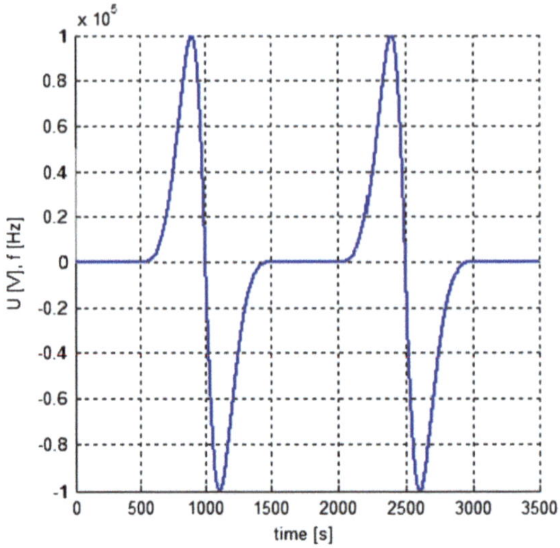

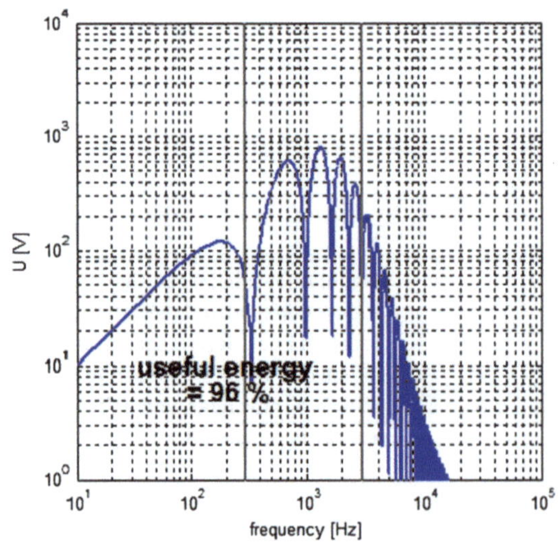

Fig. 5: Above: a signal with two balanced half-cycle chirps with a gap in between. Below: spectrum of the signal calculated with discrete FFT shown by gray lines where the signal energy is calculated. The signal retains 96 % of the energy in the desired one-decade band.

results (Figs. 5 and 6) show that, even if the half-cycle balanced chirp is used for repeated stimulation, the energy efficiency is still retained above 90 % in the desired one-decade band.

5 Conclusion

It has been found that short balanced chirp pulses (chirplets or titlets) with a duration of only a quarter or half of the full cycle concentrate about 95 % of the generated

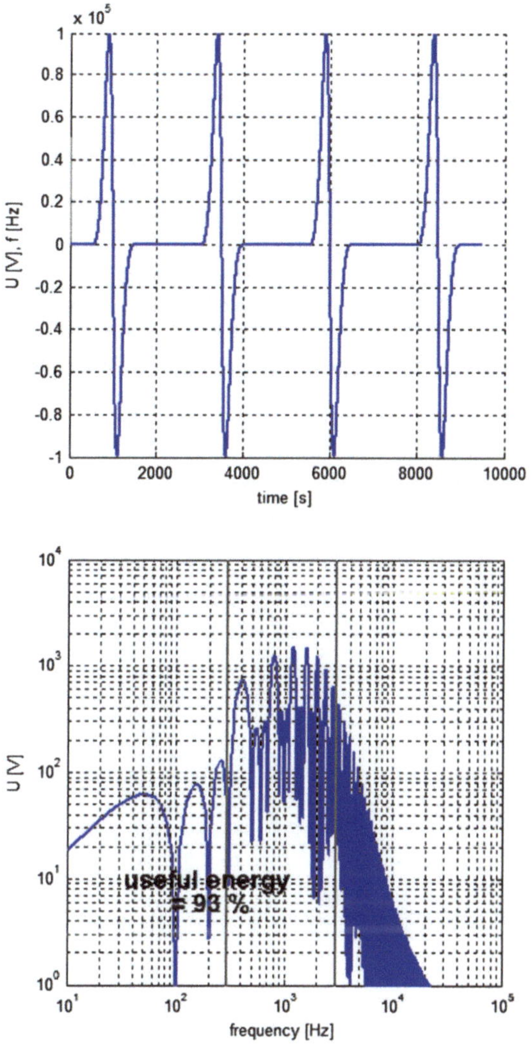

Fig. 6: Above: a signal with four balanced half-cycle chirps with a gap in between. Below: spectrum of the signal calculated with discrete FFT shown by gray lines where the signal energy is calculated. The signal retains 93 % of the energy in the desired one-decade band.

energy into the most desirable frequency range from f_1 to f_2. The optimal frequency f_P [3, 4] is the peak value of the phase spectrum of $Z(j\omega)$. The chirp-based stimulation pulses with optimized waveforms enable to concentrate the stimulation energy predominantly into the most sensitive frequency band. Short half-cycle chirps with an identical but reversed balancing part show remarkable concentration to the limited frequency band and are the preferred waveforms. Even if used for repeated stimulation, the spectrum stays limited to the desired band with above 90 % efficiency. The true effectiveness of the balanced chirp pulses for deep brain stimulation remains to be tested *in vivo*.

Bibliography

[1] A. Latteri, P. Arena, and P. Mazzone, “Characterizing deep brain stimulation effects in computationally efficient neural network models”, Nonlinear Biomedical Physics, 2011.
[2] M. K. Kuncel and M. G. Warren, “Selection of stimulus parameters for deep brain stimulation”, Clinical Neurophysiology, vol. 115, 2004.
[3] United States Patent no.: 8,788,042 B2, “Apparatus and Method for Optimized Stimulation of a Neurological Target”, Mercanzini et al. Date of Patent: Jul. 22, 2014.
[4] B. Wang and J. Weiland, “Analysis of the peak resistance frequency method for extracting tissue resistance from impedance spectroscopy”, Transactions on Biomedical Engineering TBME-00269, 2015.
[5] V. K. Khanna, “Neural stimulation and charge balancing approaches”, In: Implantable Medical Electronics, 2016.
[6] R. Gordon, M. Min, and R. Land, “Frequency-optimized rectangular pulses for deep brain stimulation”, International Journal of Bioelectromagnetism, 2015.
[7] C. R. Butson and C. C. McIntyre, “Differences among implanted pulse generator waveforms cause variations in the neural response to deep brain stimulation”, Clinical Neurophysiology, vol. 118, 2007.

Jaan Ojarand, Raul Land, Mart Min, and Marek Rist

How many frequencies to use in electrical bioimpedance measurements

Abstract: The required frequency range of electrical bioimpedance (EBI) measurements depends on the dielectric properties of the object and covers about three to four decades typically within the kHz to MHz range. The accuracy of fitting of the equivalent circuit-based impedance model with the measurement results depends on the number of frequency components used. In general, the use of more frequencies improves the fitting accuracy but complicates the measurements. However, the fitting accuracy also depends on the signal-to-noise ratio (SNR) of measurement signals. If covering the frequency range with a simultaneous multifrequency excitation, the root mean square values of its frequency components are decreasing when their number increases. It follows that the number of used frequency components has two contradictory effects on the accuracy of estimation of parameters of the equivalent circuit model. This chapter deals with the analysis of typical EBI model-fitting situations and shows how the number of frequency components influences the fitting error. Our research shows that in the case of multifrequency excitation the accuracy of fitting of the typical equivalent circuit-based EBI model may not increase when the number of frequencies is increasing. The results described here could be useful in other implementations of impedance spectroscopy.

Keywords: electrical bioimpedance, impedance measurement, impedance spectroscopy, equivalent circuit model, estimation of parameters

1 Introduction

Electrical bioimpedance (EBI) spectra are widely used to characterize the structure of tissues and cell cultures [1]. For the characterization of biological matter, wide band impedance spectroscopy is required that typically covers three to four frequency decades.

Acknowledgement: The work presented in this chapter has been supported by the institutional grant IUT19-11 of the Estonian Research Council and the European Regional Development Fund both supporting the Estonian IT Center of Research Excellence EXCITE and Competence Centre ELIKO.

Jaan Ojarand, Raul Land, Mart Min, Thomas Johann Seebeck Department of Electronics, Tallinn University of Technology, Tallinn, Estonia, e-mail: Jaan.Ojarand@ttu.ee
Marek Rist, Thomas Johann Seebeck Department of Electronics, Tallinn University of Technology, Tallinn, Estonia; and Eliko Competence Centre, Tallinn, Estonia

https://doi.org/10.1515/9783110558920-016

An important efficiency criterion of the of impedance measurements is the signal-to-noise ratio (SNR) which is proportional to the power of measurement signal. In EBI measurements, unfortunately, there is no possibility to improve the SNR by increasing the overall amplitude of the excitation signal since it is limited due to different reasons, such as non-linearity and safety requirements, as well as power consumption and some other technical issues [2, 3].

In situations where the properties of the measurement objects are changing in time, e. g., a beating heart muscle or fast moving cells in a high-throughput microfluidic channel, the frequency range of interest must be covered within a short time frame. The use of multifrequency excitations containing several frequency components simultaneously as a multisine signal does [3] gives us a possibility to perform the measurement procedure in a much shorter time span than in the case of subsequent measurements at a single hopping or sweeping frequency. We consider such cases in this chapter. However, the disadvantage is that energy of the multifrequency excitation spreads among multiple frequency components, which decreases the power of each component in the case of a limited total amplitude of the signal.

Another factor, characterizing the power limitation of individual signal components, is the crest factor (CF) of a complex excitation signal – the ratio of its peak and root mean square (RMS) values. In the current analysis, we use a multisine signal (a sum of several sine waves with different frequencies) as the multifrequency excitation. For a graspable comparison of results, some variables must be predetermined. Since the overall amplitude of signals is limited, it is reasonable to normalize the amplitude (peak value A_{pk}) of the sum of k components of the multisine $\mathrm{Max}|s(t)| = A_{pk} = 1$. In the case of equal amplitudes of k multisine components, their individual RMS values are also equal and depend only on a number of components k and the CF [3], i. e.,

$$S_{RMS}(i) = \frac{A_{pk}}{\sqrt{k} \times CF} = \frac{1}{\sqrt{k} \times CF}. \tag{1}$$

The units of signal S can be either V or A, depending on the choice to use voltage or current excitation. In Fig. 1, the essentials of formation of the EBI described in [4] and a simplified equivalent circuit of EBI are illustrated.

At lower frequencies, current passes mostly through extracellular fluid. At higher frequencies, when the reactance of the membrane capacitances becomes lower, it also passes through intracellular fluid. A more accurate electrical model also includes both, a constant phase element (CPE) acting at lower frequencies (α dispersion range) and a stray capacitance of the measurement circuitry. The required frequency range depends on dielectric properties and the structure of the sample under test (SUT) covering decades within the kHz to MHz band [1].

The accuracy of fitting of the frequency response curve with the equivalent circuit composed on the bases of measurement results depends on the number of components in the equivalent circuit, the shape of the frequency response of impedance

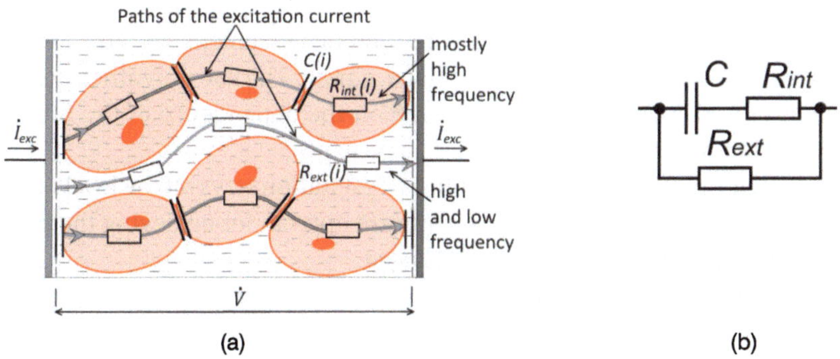

Fig. 1: (a) Essentials of formation of the EBI. (b) The three-element equivalent circuit.

and, finally, also on the number of frequency components used. The use of more frequencies improves the accuracy because the shapes of the curves of the spectra are followed better. However, the shapes of EBI spectra are relatively smooth since the elements of the equivalent circuit are not producing resonances.

Furthermore, the power of noise and its spectral properties have a substantial negative impact on the fitting accuracy.

Not many results of the systematic study of the number of required frequencies are available. Analysis of the three-element equivalent circuit obtained with sequential single-frequency excitation is described in [5] and [6].

2 Methodology

A method used for testing of the influence of the number of frequencies k on the accuracy of estimation of values of equivalent circuit elements is illustrated in Fig. 2. Computer simulation of all the parameters and the equivalent circuit at the predetermined known values of its model elements are used instead of a real SUT. In real experiments, there are more variables, e. g., parameters of electrodes, internal noises of

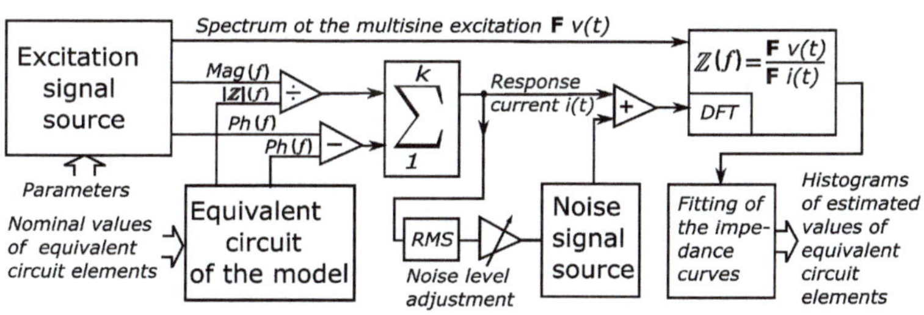

Fig. 2: A simplified structure of calculations used for testing of the influence of the number of frequencies k on the accuracy of the estimation of values of equivalent circuit elements.

the SUT, stray impedances of the connection wires and measurement circuitry. Moreover, an equivalent circuit model of real SUT is not accurate. The use of synthetic data fixes the number of variables and allows focusing on the dependence of the fitting accuracy on a number of excitation frequencies.

It is assumed that the amplitude of the multisine excitation stays on the level of ±1 V, and the initial phases of frequency components are optimal for achieving the best CF. An optimization algorithm described in [3] is used.

We have used five quasi-logarithmically distributed frequency sets in the current analysis. Three decades are covering the frequency range from 1 kHz; see Table 1. The number of frequency points per decade varies from 1 to 5. Amplitudes of all the sinusoidal components of the excitation voltage are kept equal, but their value is $V(i) = \sqrt{2} \times V_{RMS}(i)$, the value that decreases when increasing the number of components; see equation (1).

Table 1: Frequency sets of multisines and corresponding crest factors.

k	List of frequencies (kHz)	CF
3	1, 19, 401	2.43
6	1, 3, 13, 47, 167, 607	2.77
9	1, 3, 7, 19, 41, 83, 179, 379,797	2.58
12	1, 3, 5, 7, 17, 29, 47, 79, 139, 241, 421, 727	2.41
16	1, 3, 5, 7, 11, 17,23, 31, 47, 73, 109, 163, 241, 359, 523, 787	2.22

A simple three-element equivalent circuit model (Fig. 1) and a more complicated six-element equivalent circuit model (Fig. 3) containing a constant phase element (CPE) were both used.

Magnitudes of each frequency component of the response current were obtained as a ratio of the magnitudes of the excitation voltage and impedance of an equivalent circuit, as shown in Fig. 3. The phase of each component of the response current was found similarly by subtracting the phase values. A waveform of the response current

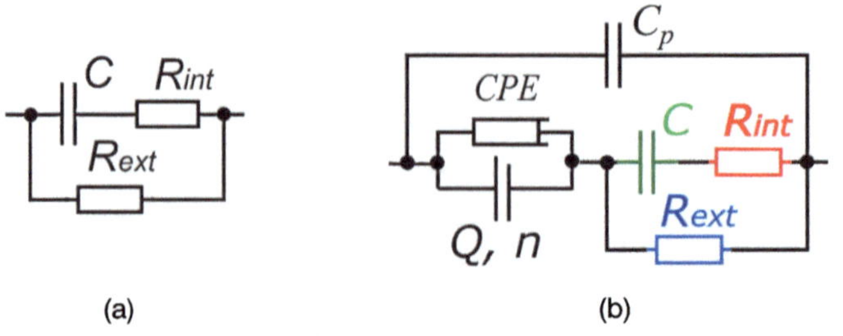

Fig. 3: (a) Three-element and (b) six-element equivalent circuit models with values of $C = 5$ nF, $Rint = 1$ kΩ and $Rext = 560$ Ω. The additionally introduced elements have values $Cp = 10$ pF, $n = 0.8$ and $Q = 10^{-6}$S.

$i(t)$, in Fig. 2, is a sum of k sinusoidal components with magnitude and phase values determined by: (a) – known multisine excitation waveform; and (b) – primarily unknown values introduced by the impedance of an equivalent circuit.

Uniformly distributed Gaussian noise adds to the response signal in the next step, and the spectrum of the summary signal is calculated using the Fourier transform (FT) with 4096 points of the waveforms. For a graspable comparison of the results, the RMS value of added noise remains constant when the number of a frequency component varies. The impedance spectrum as the ratio of spectra of excitation and response signals is calculated, and the fitting of noisy impedance curves with a frequency response of the equivalent electrical circuit is performed. Then the differences between the obtained and ideal values are calculated, and the data for histograms are collected.

The fitting algorithm minimizes the mean square error (MSE) as a relative measure of the residuals between the modeled curve values and the actual observed values. Parameters of all components of the equivalent circuit are fitted at once. Along with the residuals also the relative importance of each component of the equivalent circuit at all the frequencies is shown, which allows the user to judge the adequacy of the model and to adjust its parameters if necessary; see the example in Fig. 4. In the current experiments, however, *a priori* knowledge of the model's topography and the range of parameters simplified the task.

Finally, the errors of mean values and their standard deviations are compared for a different number of frequency components.

The SNR of the response signal is kept at the level of 10 dB, and 500 readings of the response and excitation signal spectra are used for each fitting of model parameters. Then, 500 fitting results are chosen for calculating the standard deviation of the error level for each number of frequencies.

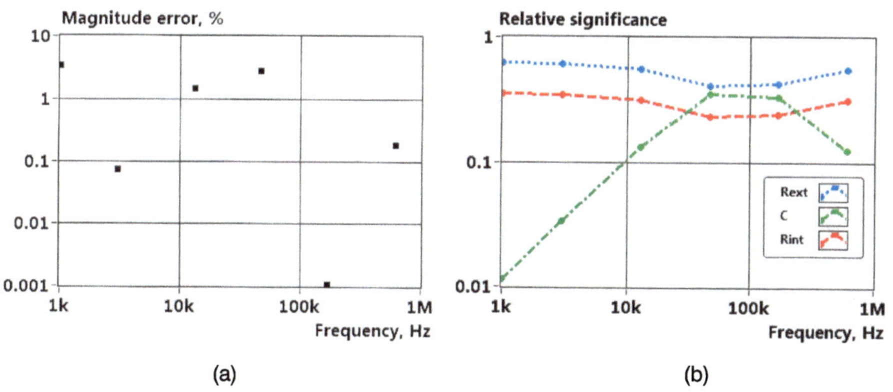

Fig. 4: (a) A snapshot of the magnitude spectrum fitting error and (b) relative significance of components in formation of the magnitude spectrum, in the case of a three-element equivalent circuit (described in Fig. 3 (a)) with multisine excitation with six components. The upper dotted line in (b) represents *Rext*, the dashed line represents *Rint* and the dash-dotted line represents *C*.

3 Results

In Figs. 5 and 6, snapshots of noisy spectra used in experiments for the fitting of values of three-element and six-element equivalent circuits are shown.The SNR of the response signal is kept at the level of 10 dB in all cases and histograms of the 500 fitting results have been collected.

In Fig. 7, histograms are shown of the 500 fitting results collected in the case of the three-element equivalent circuit for different values of C and $Rint$ using multisine excitation with three components.

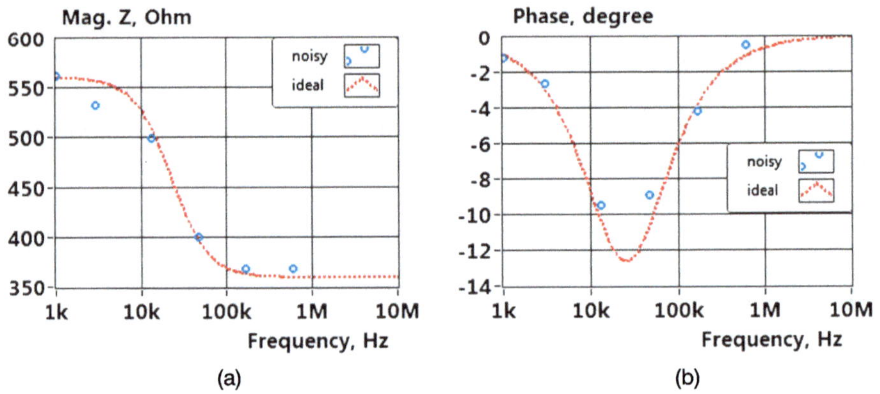

Fig. 5: (a) Snapshots of the noisy magnitude and (b) phase spectra, in the case of a three-element equivalent circuit (shown in Fig. 3 (a)) and multisine excitation with six frequency components. The dotted lines represent calculated ideal spectra and the circles denote instant values of spectra at used frequencies.

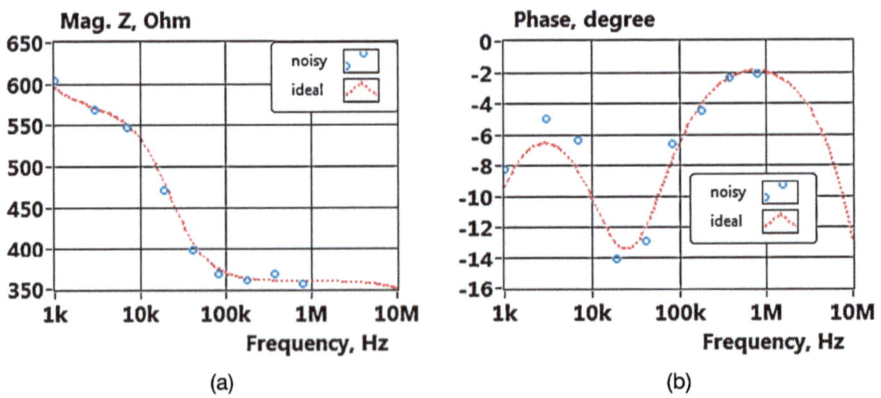

Fig. 6: (a) Snapshots of the noisy magnitude and (b) phase spectra, in the case of a six-element equivalent circuit (shown in Fig. 3 (b)) and multisine excitation with nine frequency components. The dotted lines represent calculated ideal spectra and the circles denote instant values of spectra at used frequencies.

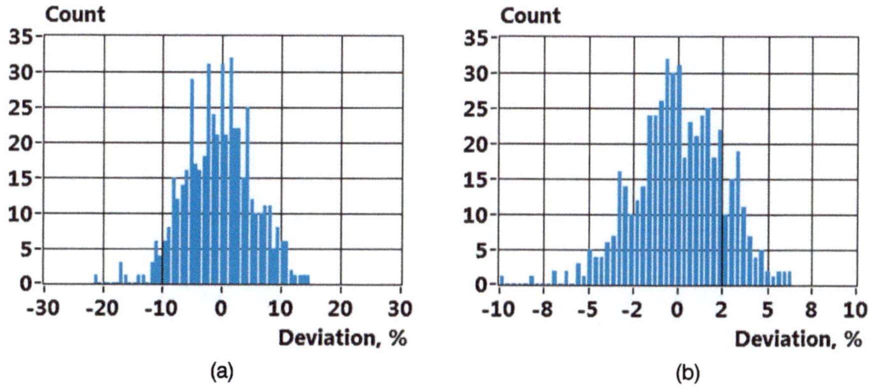

Fig. 7: Histograms of 500 fitting results of the values of (a) three-element equivalent circuit elements *C* and (b) *Rint* given for the multisine excitation having three components.

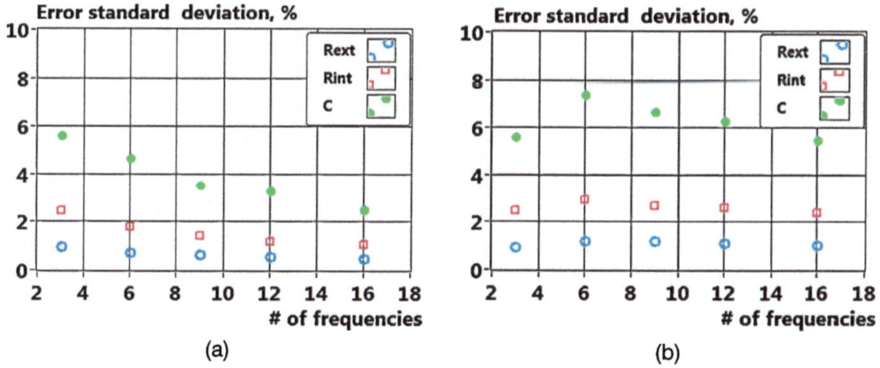

Fig. 8: The standard deviation of the fitting error of circuit element values of three-element equivalent circuit elements in the case of (a) sequential single frequency excitation and (b) multifrequency excitation; circular dots represent *C*, rectangles represent *Rint* and circles represent *Rext*.

The dependence of the standard deviation of the fitting error from the values of three-element and six-element equivalent circuit elements on the number of frequencies is illustrated correspondingly in Fig. 8 and Fig. 9.

4 Conclusion

Increasing the number of frequencies has a low impact on the accuracy of fitting results in the case of using multifrequency excitation and simple three-element equivalent circuits. Even three frequencies could provide sufficiently close results. Reduced energy of individual signal components appearing when adding the frequencies restrains the effect of using a larger number of frequencies.

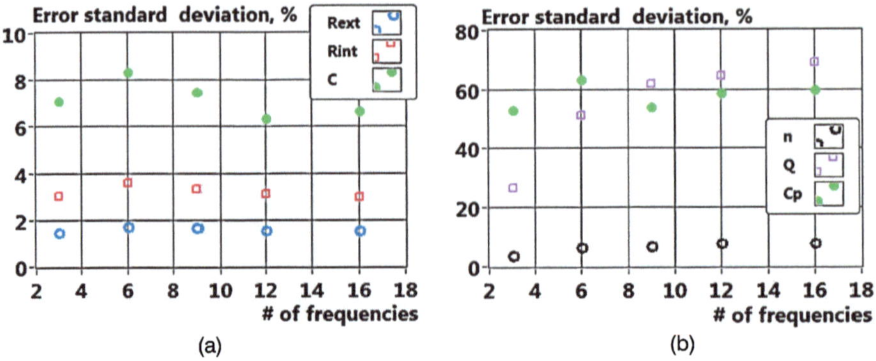

Fig. 9: Standard deviation of the fitting error of the values of elements of six-element equivalent circuit in the case of (a) multifrequency excitation and (b) a corresponding equivalent circuit. Values of *C*, *Rint* and *Rext* are the same as of the three-element circuit in Fig. 1, but the values of additionally introduced elements are Cp = 10 pF, $n = 0.8$, and $Q = 10^{-6}$.

It is reasonable to keep the number of frequencies as low as possible at the acceptable errors. Expectedly, the number of frequencies should be larger, if the noise distributes non-uniformly over the frequencies or some concentrated disturbances appear at some frequency sub-ranges. Every specific case deserves special handling.

The positive effect from the use of more frequencies is larger in the highly noisy environment if the single-frequency signals are applied sequentially for excitation, but then the measurement process becomes too long for the measurements of time-varying impedances.

Bibliography

[1] S. Grimnes and Ø. G. Martinsen, "Bioimpedance and Bio-electricity Basics", 3rd ed., Elsevier-Academic Press, pp. 77–403, 2015.
[2] U. Pliquett, "Electricity and biology", In: Proc. BEC2008, Tallinn, Estonia, pp. 11–20, 2008.
[3] J. Ojarand and M. Min, "Crest factor optimization of the multisine waveform for bioimpedance spectroscopy", Physiological Measurement, vol. 35, pp. 1019–1033, 2014.
[4] M. Min, T. Parve, and U. Pliquett, "Impedance detection", In: Encyclopedia of Microfluidics and Nanofluidics, 2nd ed., (Ed.) Prof. Dongqing Li, Springer, Berlin–Heidelberg, 28 pp., 2015.
[5] L. Ward and B. Cornish, "Multiple frequency bioelectrical impedance analysis, how many frequencies to use ?", In: Proc. ICEBI XII and EIT V, vol. 61, pp. 321–324, 2004.
[6] S. Kun and R. A. Peura, "Selection of measurement frequencies for optimal extraction of tissue impedance model parameters", Medical & Biological Engineering & Computing, vol. 37, pp. 699–703, 1999.

Andres Kink, Marek Rist, Raul Land, Hip Kõiv, and Mart Min

A method of intra-ventricular bioimpedance spectroscopy to estimate the dynamic volume of right ventricle

Abstract: Dynamic volume measurements of the heart are primarily done in the left ventricle, but as the right ventricle is a frequent place for pacing electrodes it could be easier to estimate cardiac output through the right ventricle. Our goal was to find the correlation between intra-cardiac bioimpedance deviations and the variations related to changes of the right ventricle volume. The variations of right ventricle volume have been evaluated experimentally using a setup with an isolated pig heart. Measurements were made with an intra-ventricular catheter and using multifrequency impedance spectroscopy.

Keywords: bioimpedance, cardiac output, right ventricle volume, estimation

1 Introduction

In Europe, 45 % of all deaths are a result of cardiovascular disease, and the costs that derive from heart failure exceed the costs of all other diseases together [1]. Therefore, the motivation to develop a method to improve diagnostics and prevent heart-related problems is considerable. The importance of dynamic volume changes and the function of the right ventricle have increased in the last decade because they provide necessary information to assess cardiovascular performance [2]. Measuring blood volume changes in the right ventricle is essential to estimate total cardiac output and it is an important parameter to characterize the heart and its pump function in experimental studies. Cardiac output is usually measured from the left ventricle [3] because its geometry is relatively simple – a sphere – while the right ventricle is like an unequal layer around left ventricle in 3D space. Also, the right ventricle is thin-walled compared to the left ventricle, making the measurements difficult and more sensitive to changes from outside the right ventricle (Fig. 1).

Acknowledgement: This research was supported by the Estonian Research Agency grant IUT1911 and the European Regional Development Fund in frames of the Estonian Center of Research Excellence EXCITE. The authors thank SMARTIMPLANT Ltd.

Andres Kink, SMARTIMPLANT Ltd., Tallinn, Estonia, e-mail: andres@linking.ee
Marek Rist, Raul Land, Hip Kõiv, Mart Min, Th. J. Seebeck Department of Electronics, Tallinn University of Technology, Tallinn, Estonia

https://doi.org/10.1515/9783110558920-017

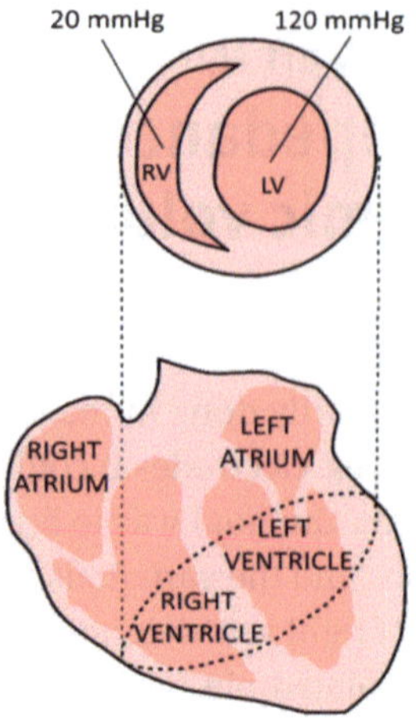

Fig. 1: Structure and shape of the adult right and left ventricles.

Although the left ventricle has a better shape for measurements, in real life the right ventricle is a basis for artificial pacing and a common place for pacing electrodes, thus making cardiac output estimation in the right ventricle potentially more convenient. Modern heart assisting devices like pacemakers are "smart devices" that can regulate the heart according to its condition. To help in the management of the unstable patient, measurements of cardiac output need to be taken repetitively [4]. Pacing electrodes are inside the right ventricle permanently, hence making constant measurements possible. Our main goal was to estimate the absolute or relative right ventricle volume changes experimentally with impedance spectroscopy and a multielectrode intra-ventricular catheter using a setup with an isolated pig heart. Multiple hypotheses have been developed and evaluated.

1) Bioimpedance can be used for the measurement of the amount of blood in the vicinity of the electrodes, however due to unknown position and shape (bending) of the electrode the measured result may not correlate with the actual volume change in the whole ventricle.

2) By using bioimpedance spectroscopy, the distance between the electrode and the cardiac tissue could be evaluated based on the changes in β dispersion area, since impedance of blood has negligible reactive components in the used frequency range from 1 kHz up to 349 kHz and heart muscle has a much larger reactive component due to the dense packing of muscle cells [5].

3) With compensation based on the estimated position of the electrode, the segment-based volume changes could be measured with better precision.

The use of a multielectrode intra-ventricular catheter for impedance spectroscopy is a potentially useful technique for repetitive measurements to estimate the right ventricular volume.

2 Methods and general principles of design

We evaluated the right ventricle volume during multiple working cycles with different loading conditions. Usually intra-cardiac impedance is measured with single frequency but we believe that by using multifrequency measurements, additional information could be gained. The change of volume is measured by inserting an Inquiry 1110-5-25-M decapolar electrophysiology catheter, shown in Fig. 3, from St. Jude Medical into the right ventricle and monitoring the change in impedance between the electrodes using the four-wire connection method. The electrode positioning is uneven: 2–5 mm, giving four 2-5-2 (distances between electrodes in mm) measurement combinations or seven when we also include 5-2-5 combinations. Measured current and voltage values from 2-5-2 and 5-2-5 need separate correlation parameters when translating measured impedance data to blood volume (Fig. 2).

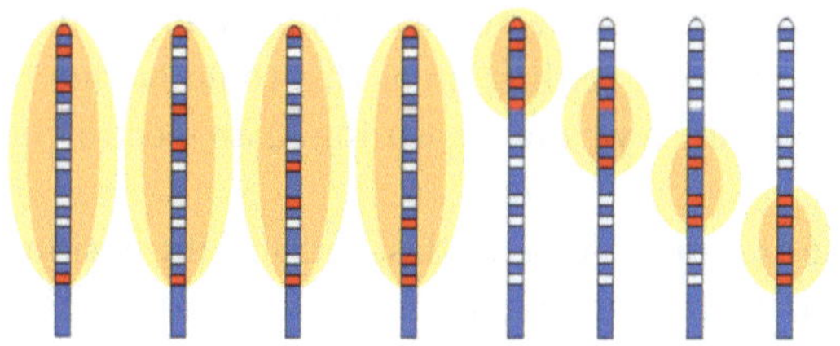

Fig. 2: Switching between electrode modes.

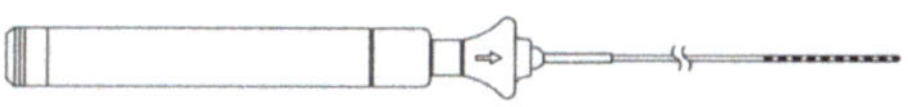

Fig. 3: Catheter that is inserted into the right ventricle through the pulmonary artery to measure impedance change (INQUIRY™ electrophysiology catheter).

The relationship between volume (V), electrode distance, impedance and blood resistivity is given as follows:

$$V = \frac{1}{\alpha} \cdot \rho \cdot L^2 \cdot (G_x - G_p), \tag{1}$$

where ρ is the resistivity of the blood, L is the distance between measurement electrodes, $\alpha = \frac{SV_{\text{conductance}}}{SV_{\text{reference}}}$ is a stroke volume (SV) correction factor, G_x is a total measured conductance and G_p is the conductance of heart muscle tissue [3]. This equation is important for left ventricle volume measurements but research [6] has shown it is accurate also in the right ventricle. Our experiments show that it is not reliable to use the given equation inside the right ventricle due to the unknown electrode position. In order to obtain information about the shape and volume of the ventricle during operation of the heart, the measurement must be made relatively fast. When using a single frequency, fast measurement is relatively easy to achieve. When using spectrum measurement, obtaining real-time measurements becomes more difficult. We have developed a real-time impedance spectrum analysis device, Quadra, which uses binary multifrequency signals for excitation and measures 15 frequency points, from 1 kHz to 349 kHz, with a period of 1 ms, giving 1000 measurements per cardiac cycle when the heart rate is 60 bpm [7].

Excitation in the form of voltage was used with V_{RMS} of 200 mV resulting in root mean square (RMS) currents in the range of 0.8 mA to 1.2 mA. Since we need to analyze four or seven sets of electrodes, a fast switching of the measurement device to electrode sets is needed. This could be done with 28 relays or one integrated cross-point switch. The best results were achieved by using jumping excitation along with measurement. Changing excitation is related to longer stabilization time, thus reducing the measurement rate.

We noticed a large measurement noise at frequencies < 10 kHz when multiplexing both excitation and sense electrodes. The noise decreases exponentially in time and can be considered negligible after 100 ms of settling time. This could be due to the formation of an electrical double layer since we are measuring in an electrolytic environment [8].

When the excitation was fixed to the first and last electrodes, and only the position of the sense electrodes was multiplexed, the measurement noise from lower frequencies was removed. The current distribution, in an already difficult environment, gets a lot more complex and interpretation of measured impedance would be more difficult.

For the study of our results we use custom-made analysis software made in LabVIEW. It shows impedance magnitude values, the pressure value and the phase angle in real time.

3 Measurement setup

For experiments, we obtained a pig's heart from the slaughterhouse, which was in the condition to be resuscitated. The electrophysiology catheter was placed inside the right ventricle through the pulmonary artery together with a silicone tube to manipulate the loading condition and the blood volume. In addition, a tube for the pressure sensor was added to measure the blood pressure in the right ventricle. The tricuspid valve was then closed to prevent the tubes from moving too much in the ventricle. The coronary arteries were kept connected for the external blood supply to keep the heart alive. Oxygenated warm blood was used for heart perfusion, since living heart needs oxygenated blood of a correct temperature (around 37°C) (Fig. 4).

For measurements, an intra-ventricular catheter was attached to the real-time impedance spectrum device which was connected to the PC and analysis software made in LabVIEW (Fig. 5). The right ventricle was then filled with blood and the heart

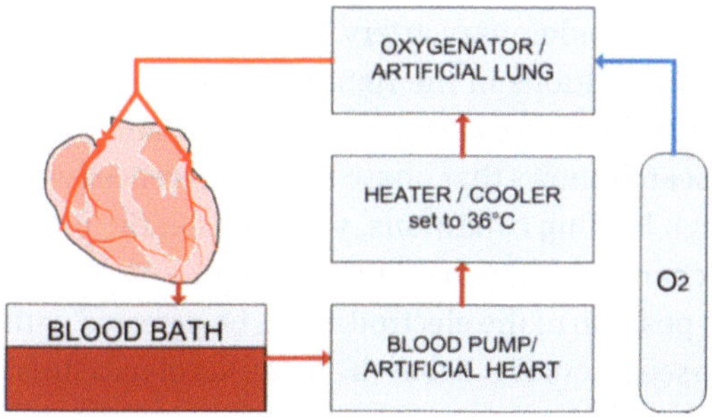

Fig. 4: Isolated retrogade perfusion heart setup.

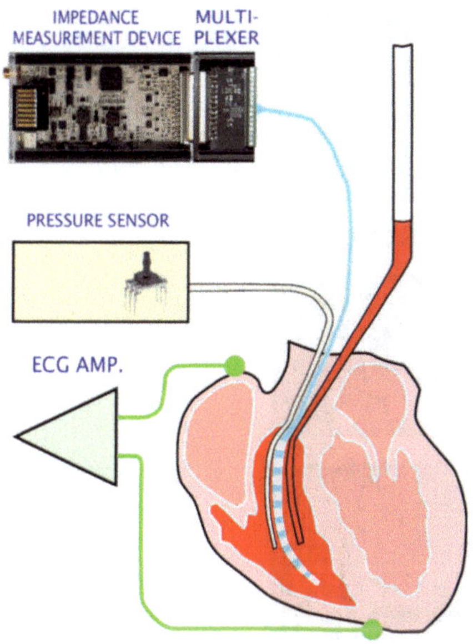

Fig. 5: Measurement setup with pressure sensor and impedance measurement device. ECG is useful to monitor heart electrical activity.

was given an electrical impulse or was contracted by manual squeezing to revive it. Changing the blood volume inside the right ventricle gave us impedance results we could expect. We used four-wire measurement. Method A consists of switching the excitation position and switching the sense position, which gives us a more uniform and localized current distribution but also more measurement noise. Method B consists of a fixed excitation position and switching the sense position, giving a shorter acquisition time but a wider and less uniform current distribution.

4 Results

A versatile, compact and portable laboratory was constructed for isolated experiments together with measurement devices. The isolated heart filling was modulated according to the level in the tube inserted into the pulmonary artery. The pulmonary valve was removed to adapt different loading conditions in the right ventricle. Under low loading conditions, prominent changes in bioimpedance were registered and evaluated in tissue distance changes. Our research shows that phase angle differences were the most noticeable. Under mean or high loading conditions, variations in the phase angle do not correlate well with value changes inside the right ventricle.

The initial position and changes in position of the electrodes can be detected with bioimpedance measurements. It can be seen from Fig. 6 that the modulus of measured

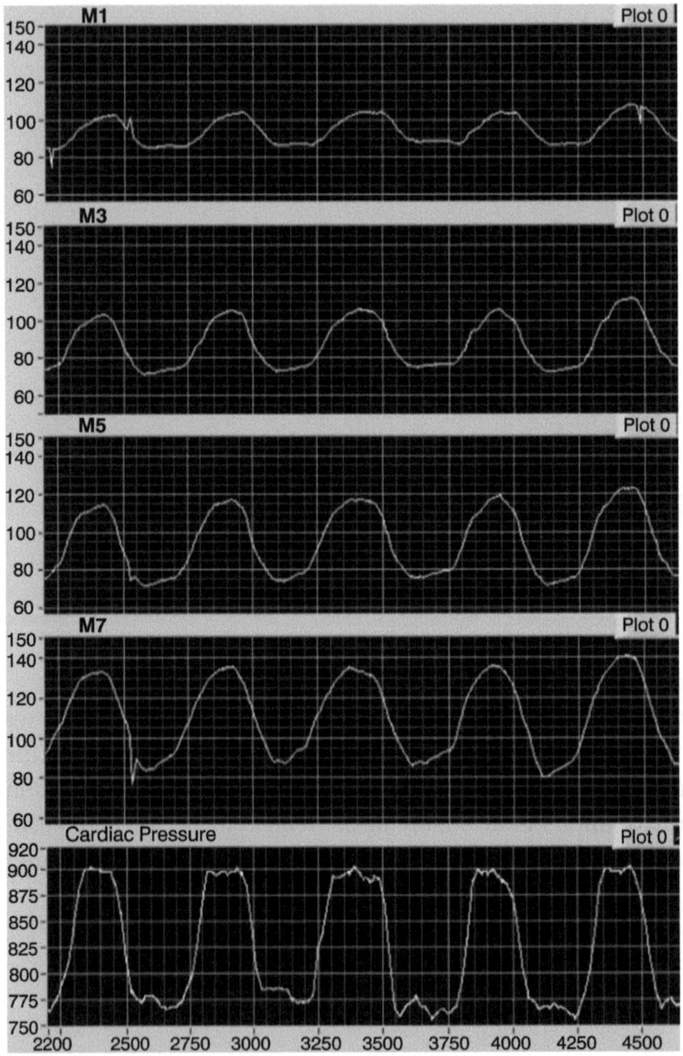

Fig. 6: Magnitude (M1, M3, M5, M7 in Ohms) and pressure (in mmHg) curves measured from the right ventricle.

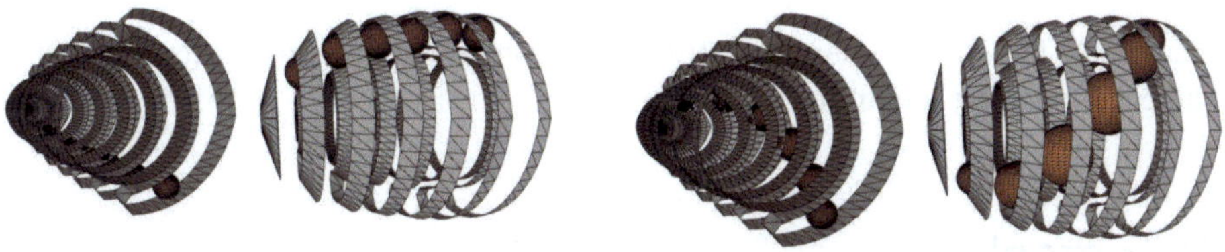

Fig. 7: Electrode positions 1 and 2 with sensitivity spaces (spheres).

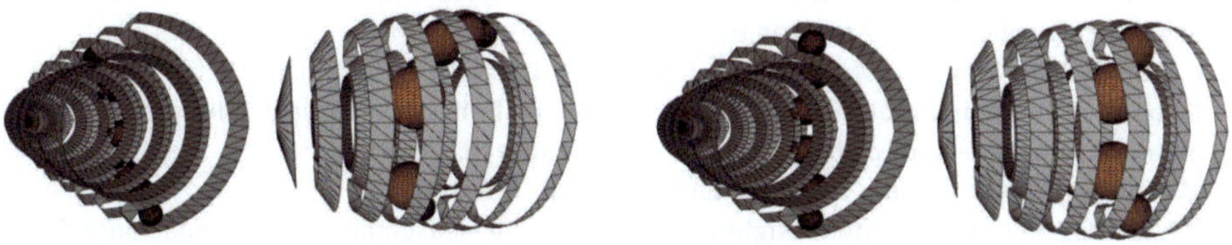

Fig. 8: Electrode positions 3 and 4 with sensitivity spaces (spheres).

impedance shows good correlation with the cardiac pressure. The changes in magnitude of different segments represent the volume of blood and the proximity of the myocardium. We are able to assess segment-based volume changes in the right ventricle. Fig. 7 and Fig. 8 show segments of the heart and how electrode positions can change during the heart cycle. Four different positions are proposed with spheres marked as electrode sensitivity spaces.

5 Discussion

In the present study, we have examined blood volume changes in the right ventricle using impedance spectroscopy and a decapolar catheter. We measured electrode shape changes, segment-based volume changes and the distance between the measurement electrode and the cardiac muscle tissue in the right ventricle by using bioimpedance measurements with an intra-ventricular electrophysiological catheter. The proposed technique is affected by the complex geometry of the right ventricle and the conductivity of surrounding tissues. Using a Baan catheter [3] there is an assumption that the electric field produced by the catheter is homogeneous, parallel to the long axis of the ventricle, and does not extend to other cardiac chambers. If these presumptions are not considered, the errors of ventricle volume estimation are significant. The outcome shows a noticeable correlation between the changes of bioimpedance and the variations of right ventricle volume, but the interpretation of results can be quite complicated due to the different electrode positions inside the right ventricle. For acceptable results, it is necessary to estimate the electrode configuration inside the right ventricle and separate impedance change contributions from blood and cardiac tissue. This is a subject of ongoing research.

Bibliography

[1] ESC, "European cardiovascular disease statistics 2017 edition", 2017.
[2] W. H. W. Tang and W. Tong, "Measuring impedance in congestive heart failure: Current options and clinical applications", American Heart Journal, 2009.
[3] J. Baan, T. Aouw Jong, P. L. M. Kerkhof, R. J. Moene, A. D. Van Dijk, E. T. Van Der Velde, and J. Koops, "Continuous stroke volume and cardiac output from intra-ventricular dimensions obtained with impedance catheter", Cardiovascular Research, 1981.
[4] M. R. Pinsky, "Why measure cardiac output?", Critical Care, 2003.
[5] P. A. White and A. N. Redington, "Right ventricular volume measurement: Can Conductance do it better?", Physiological Measurement, 2000.
[6] H. Helmholtz, "Ueber einige Gesetze der Vertheilung elektrischer Ströme in körperlichen Leitern mit Anwendung auf die thierisch-elektrischen Versuche", Annalen der Physik und Chemie, 1853.
[7] M. Rist, M. Reidla, R. Land, T. Parve, O. Märtens, P. Annus, J. Ojarand, and M. Min, "Modular system for spectral analysis of time-variant impedances", In: IFMBE Proceedings, 2013.
[8] S. Grimnes and Ø. G. Martinsen, "Bioimpedance and Bioelectricity Basics", Academic Press, 2000.

Mahdi Guermazi, Olfa Kanoun, and Nabil Derbel

Meat quality monitoring by classification based on impedance spectra

Abstract: In this contribution, we investigate the possibilities of meat state assessment based on bioimpedance spectroscopy and fuzzy methods. We report on the procedure of feature selection from impedance spectra of meat aiming to reduce measurement data to features providing sufficiently information to build the basis for meat state estimation, classification of meat type and estimation of the level of freshness. The developed method is suitable for signal processing of impedance spectra in a devices for real-time meat monitoring based on bioimpedance spectra.

Keywords: impedance spectroscopy, bioimpedance, fuzzy logic, feature extraction, data mining

1 Introduction

Meat quality cannot be measured directly, like a classical measurement quantity. It includes different aspects related to microbial attacks, the existence of certain substances and hormones and structural changes of the muscle during maturation processes. Especially for beef, maturation and structural muscle changes are important to realize a tender state by consumption. In order to follow muscle strutural changes, electrical measurement methods, such as bioimpedance. Bioimpedance spectroscopy is fast, non destructive and can be realized at low-cost.

Several investigations have been carried out in the field of meat and food diagnosis. Different features of the impedance spectra can be adopted for data mining such as the use of model parameters, the use of characteristic points, the extraction of data at a fixed frequency or the use of the whole measurement results. An overview is given about different results realized with data mining and classification methods related to the topic of biological tissue monitoring.

Guerrero et al. [1] have used Cole-Cole model parameters to describe the relationship between sensory properties and electrical parameters of different meat qualities of hams for the semimembranosus (SM) and biceps femoris (BF) muscles. The used parameters are R_0, which corresponds to the electrical impedance at lower frequencies,

Mahdi Guermazi, Chair for Measurement and Sensor Technology, Chemnitz University of Technology, Reichenhainerstrasse 70, 09126 Chemnitz, Germany; and National School of Engineers of Sfax, University of Sfax, Sfax, Tunisia, e-mail: mahdienis@yahoo.fr
Olfa Kanoun, Chair for Measurement and Sensor Technology, Chemnitz University of Technology, Reichenhainerstrasse 70, 09126 Chemnitz, Germany
Nabil Derbel, National School of Engineers of Sfax, University of Sfax, Sfax, Tunisia

https://doi.org/10.1515/9783110558920-018

R_{inf}, which corresponds to the electrical impedance at high frequencies, α, which corresponds to the shape adjustment parameter, and f_c, which corresponds to the characteristic frequency of the region under measurement at which the imaginary part of the electrical impedance has the largest absolute value. A fifth parameter is introduced: the ratio between R_{inf} and R_0, which is proportional to the ratio of extracellular water to total water in meat. Principal component analysis is used in order to group different hams and to locate them in the biplot. They are classified in three groups according to the main components: hams having pH45 (defined as the pH at 45 min postmortem) lower than 5.85 were classified as pale, soft and exudative (PSE), hams having pHu (defined as the pH at 24 h postmortem) higher than 6.0 as dark, firm and dry (DFD) and the rest as normal hams. The electrochemical impedance spectroscopy (EIS) prototype correctly detected 69.2 % and 56.0 % for SM and BF muscles, respectively, of the problem hams in terms of pastiness. We suggest that the electrical parameters evaluated in green hams by the EIS prototype could be useful for predicting pastiness in dry-cured ham.

Also, for the application of bovine tissue classification, Negri et al. [2] have used features for classification composed by the Cole-Cole parameters R_0, R∞, τ and α. Results show that using the fitted Cole-Cole model parameters instead of the full spectrum as the input of a neural network can enhance its classification rate and significantly reduce its topology when tested with experimental bovine tissue spectra.

The same method has been used by Filho et al. [3] for classification to assess bovine milk quality. They have confirmed the use of the Cole-Cole parameters may have a better classification than the use of the raw acquired spectrum. Using a multilayer perceptron (MLP), the recognition percentage can reach 94.6 % to classify adulterated milk samples with the type of impurity added being water (H_2O) or hydrogen peroxide (H_2O_2). The used MLP is composed by a hidden layer with two neurons, an output layer with three neurons defining each class (milk, milk with H_2O, milk with H_2O_2).

Meat monitoring is done using the three main parts: the experimental investigations, the modeling and features analysis and finally the meat type and freshness diagnosis (Fig. 1).

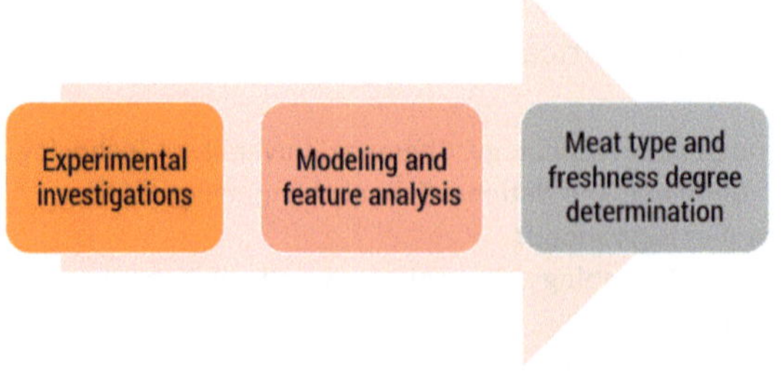

Fig. 1: Monitoring process of meat quality.

2 Methodology

The most important and necessary information from a meat monitoring device is the type of the muscle and the freshness of the meat. To ralize a suitable signal processing, it is necessary to remove spikes in the measurements due to errors. Another important factor that should be respected is the use of a reduced input in order to reduce the effort and increase the stability of implementation. The process of classification is therefore composed of two principal steps:
- Accurate feature selection from the measured impedance spectra;
- Classification with high reliability.

2.1 Feature selection

Before classfication, generally feature selection should be carried out to reduce the measured data to specific data, which serve as the basis for classification. Selection of features represents therefore a decisive step in order to reach a high classification probability during learning and test phases. Measurements have been carried out with different tissues and for a long period of time in order to monitor the degradation state with time. The measurement setup consists of a laboratory impedance analyzer (Agilent 4294 A) connected to a personal computer for data acquisition. The spectrometer generated a signal within a frequency band ranging from 40 Hz to 110 MHz (Fig. 2). Cylindrical penetrating multielectrodes are used for measurements ensuring the reduction of the anisotropy effect and good contacts between electrodes and the tissue [4].

There are different possibilities for selecting features from impedance spectra, such as the use of model parameters, the use of characteristic points, the extraction

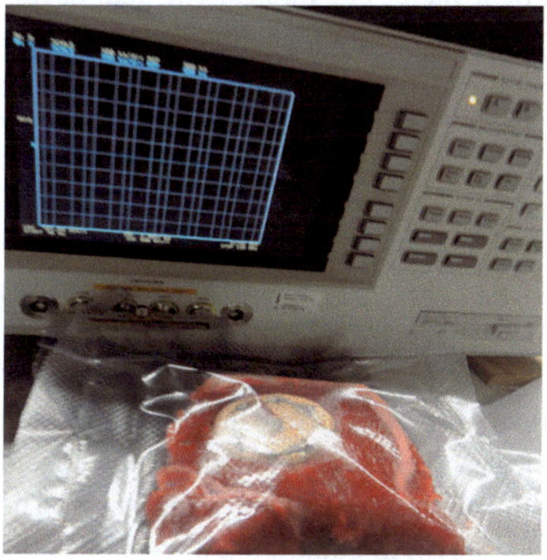

Fig. 2: Meat measurement.

of data at a fixed frequency or the use of the whole measurement results. According to literature and to experimental investigations, measurement results have been evaluated by means of the modified Fricke-Cole-Cole model. Results are reproducible and correspond to the expected behavior due to aging [5]. The model parameter evolution shows principally the same behavior, but differs from one muscle to another in the absolute value. This is due to several factors, especially the biologic characteristics of each muscle.

Both model parameters (MP) and characteristic points (CP) can be principally used for data mining and represent an accurate feature gathered from the measured impedance. The investigated possibilities are MP, CP and MP + CP (Fig. 3).

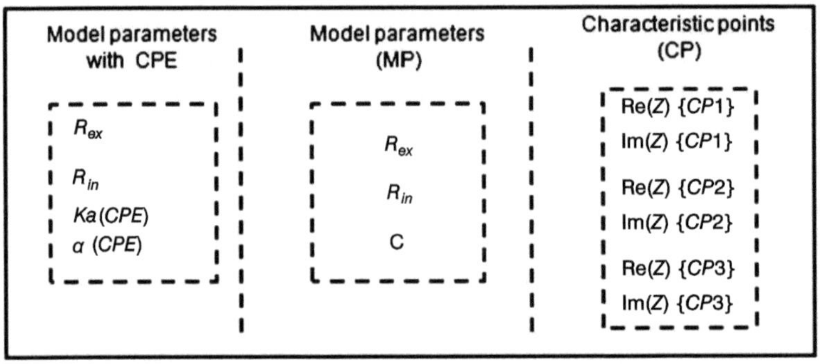

Fig. 3: Selected examples for feature extraction from impedance spectra.

According to the literature, principal component analysis (PCA) has been often used to reduce features to only the necessary inputs [6–8]. PCA belongs to statistical methods of multivariate analysis [9]. It is a method of transforming the initial dataset of high dimensions into a new dataset represented by a vector of samples. The goal of this transformation is to concentrate the information and reduce its dimension by considering the data variance. The PCA method is a linear transformation, as described in equation (1), which transfers the original possibly corresponding data set X to a new uncorrelated dataset F_{new}. The new dataset F_{new} has the same dimension as the original dataset X; A_t is the coefficient transformation matrix. We have

$$F_{\text{new}} = A_t X. \tag{1}$$

The first few components (columns) of the new dataset F_{new} contribute the majority of variances and are considered to be the principal components, which can reflect the original data matrix without big information losses. They have smaller dimension and are uncorrelated to each other, which makes the analysis of the data much easier. The input data set X_d is in the form of a matrix, as shown in equation (2), where p is the number of the parameters and n is the number of the observations. We have

$$X_d = \begin{pmatrix} X_{11} & X_{12} & X_{1p} \\ X_{21} & X_{22} & X_{2p} \\ X_{n1} & X_{n2} & X_{np} \end{pmatrix}. \tag{2}$$

The first two components for the feature composed by MP, for the CP and for the MP + CP explain more than 90 % of the variances. In order to verify which feature we should use for classification, a simple idea is used consisting of the principle of superposition of the test data in the corresponding region of input data after k-nearest neighbors (KNN) calculation. The algorithmic method of KNN was used to give a complete 2D presentation in a geometrical plot of the data.

2.2 Classification method

Different suitable methods of classification can be used in this application. Fuzzy logic is selected as a classification method that provides a successful alternative for easy classification, as it is in general a robust method. It has also the advantages to deal with uncertainty problems and usually gives good results for various areas. Fuzzy logic is widely used in medical applications for diagnosis of tuberculosis, cancer, asthma, diabetes, aphasia, malaria, HIV, pulmonary embolism, cortical malformations and pancreatic diseases [10]. Fuzzy logic has been implemented for an earlier detection of breast cancer [11, 12]. Results are found to be very useful and helpful for oncologists, radiologists and doctors. Thus, fuzzy logic, as an intelligent method, will be of great help especially when saving patient lives is possible. The process of fuzzy logic based classification includes three principal steps [13]:

- Fuzzification of input variables and choice of suitable membership functions;
- Construction of fuzzy rules and making decisions;
- Defuzzification.

An explanatory illustration of the fuzzy inference process is shown in Fig. 4.

In fact, the model parameters (received data) undergo a feature's extraction process (information), which is a preparatory step for the classification method. The resulting data are then fuzzified using membership functions. Afterward, fuzzy values are analyzed using fuzzy rules to get the fuzzy outputs. After defuzzification, the new data are then compared and classified with similarities to get the desired clusters.

Fuzzification is the conversion of crisp input data into fuzzy values or linguistic variables using membership functions. These fuzzy values reflect the human perspective of the given system. These fuzzifiers are membership functions determined depending on a subjective view of the problem and an individual's perception of the situation. Depending on the desired class and performances, simple functions are used to build membership functions, such as triangular, trapezoidal and Gaussian ones.

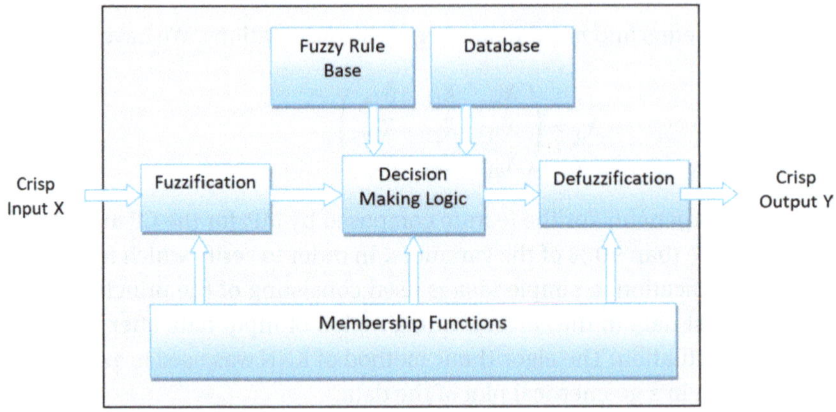

Fig. 4: Block structure of a Fuzzy system.

Fuzzy rules, which have the general form of "IF X is A THEN Y is B", are destined to map the fuzzy input X to the fuzzy output Y. In a logical context, X is referred to be a linguistic variable and A is a linguistic value. The 'IF' part of the rule is called antecedent; the 'THEN' part is called consequent. Connection between the rules is achieved using logical operators AND, OR and NOT. AND or min takes always the minimum of the functions value. OR (max) takes the maximum of the presented functions.

Defuzzification is the conversion of a fuzzy set quantity to a precise and deterministic set quantity. This step is necessary because in several applications, where the implementation is required, machines could only work with crisp or binary values and not with linguistic variables.

After defuzzification, the new data are compared and classified to get desired clusters [14]. Two important types of fuzzy logic systems exist and are much used for diagnosis: Mamdani inference [15, 16] and Takagi–Sugeno (TS) inference [17].

The Mamdani method is suitable for nonlinear systems, characterized by a fuzzy set consequent in the fuzzy rule that has the following form:
R : IF L(x_1 is A_1,....,s_k is A_k) THEN (y_1 is B_1,....,y_n is B_k),
where:

- x_1,.., x_k are the inputs of membership functions;
- A_1,.., A_k are the fuzzy sets of the antecedent part;
- y_1,.., y_n are the outputs of the membership functions;
- $B_1, \ldots, B_n$ are the fuzzy sets of the consequent part and L is the logical function connecting the propositions.

For the TS method, the output membership function or more precisely the rule consequent is only linear or constant. If the rule consequent is constant we call the model a zero-order Sugeno model, which has the following form:
R_0 : IF L(x_1 is $A_1, \ldots, s_k$ is A_k) THEN $y = k$,
where k is a crispy constant defined by the operator.

If the rule consequent is a linear function the Sugeno model is called a first-order Sugeno model with the following expression:
R_1 : IF L(x_1 is A_1,...,s_k is A_k) THEN $y = f(x_1,...x_n)$.

3 Results and discussion

3.1 PCA computing

A database composed of six muscles is used in order to use the PCA method well. This database will serve to determine the suitable feature. Three different muscles of two animals of different ages are used. The test data should be unknown. For that, two spectra will be used for learning and two others for testing in every case.

Singular-value decomposition is chosen following the study done in [18]. The decomposition of the normalized matrix "X" is done according to

$$X'' = \mathrm{USV}', \tag{3}$$

where S is the $n \times p$ diagonal matrix with singular values σ_i of the matrix X in diagonal form. We also have

$$S_{ii} = \sigma_i \tag{4}$$

and singular values are already sorted in a decreasing order, i. e.,

$$\sigma_1 > \sigma_2 > \sigma_2 >> \sigma_p. \tag{5}$$

The relationship between the singular value σ_i and the eigenvalue λ is given by

$$\sigma_i = \sqrt{\lambda_i}, \tag{6}$$

where U is the $n \times n$ left singular-valued vector matrix of X, V is the $p \times p$ right singular-valued matrix of X and both are orthogonal matrices. The principal components F can be calculated by equation (7), i. e.,

$$F = X.V = U.S.V'.V = U.S. \tag{7}$$

Before dealing with PCA computing, the inputs and test data of the three cases, MP, CP and MP + CP are defined. The test data have exactly the same dimension as the input data.

The result of the superposition of model parameter test data on the KNN partition corresponds to an accuracy equal to 100 %, as shown in Fig. 5. The accuracy for the characteristic points is 66 % with four erroneous outputs. For the feature composed

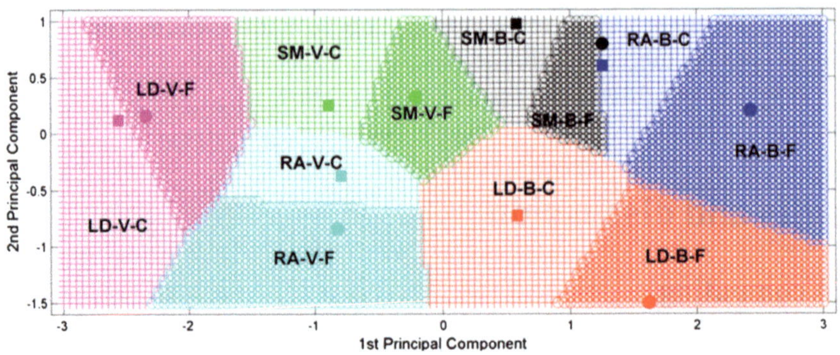

Fig. 5: Superposition of model parameter test inputs to the KNN graph.

by the characteristics points and model parameters we get 12 correct outputs that correspond to an accuracy of 100 %.

Feature extraction based on the use of model parameters and characteristic points shows good results but with the consideration of the use of reduced inputs in the network we choose to work with the model parameters only for the classification that also gives an accuracy of 100 %.

3.2 Zero-order Sugeno model

The classification is carried out using the TS method, which is similar to the Mamdani model in fuzzifying the inputs and applying the fuzzy operator. The major reason of this choice is that the TS method output membership can be either constant (zero-order) or linear (first-order), corresponding exactly to the application specifications. The zero-order TS method with constant output memberships is exactly the solution that permits to get an accurate diagnosis and that corresponds to the problematic. For each muscle we appoint a corresponding constant number:

- 1 for the muscle LD beef (M1);
- 2 for the muscle RA beef (M2);
- 3 for the muscle SM veal (M3);
- 4 for the muscle LD veal (M4).

And for each muscle we appoint a corresponding constant number as output:

- 1 for fresh meat (F);
- 2 for edible meat (E);
- 3 for critical meat (C);
- 4 for dangerous meat (D).

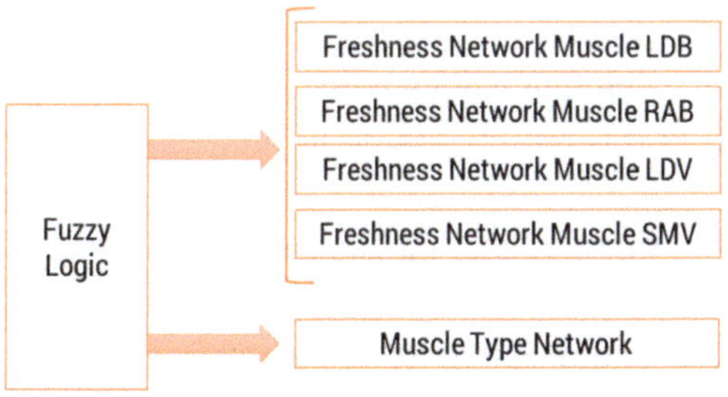

Fig. 6: Fuzzy logic process.

The fuzzy logic process (Fig. 6) is composed of two fuzzy logic systems that have been used for meat diagnosis; the first for the classification of the muscle type, the second for finding the freshness, to classify the meat in four classes that correspond to four different periods of time: fresh meat, edible meat, critical meat and dangerous meat.

The output membership function or more precisely the rule consequent is only constant. We have
R_0 : IF L(x_1 is $A_1, \ldots, s_k$ is A_k) THEN $y = k$.

3.2.1 Membership functions

The first step of the muscle type classification consists of the choice of the input and output variables and selecting carefully their intervals' ranges. By this step, we try to collect the data which generate the same output in one interval range. This process is carried out for all the inputs. In one side, we define the three model parameters R_e, R_i and C, extracted from bioimpedance measurements as the inputs [5]; the output corresponds to the muscles LD_{Beef}, RA_{Beef}, SM_{Veal} and LD_{Veal}. After several trials for the muscle type classification, we select the Gaussian membership functions (MF) defined by a central value c and a standard deviation $\sigma > 0$. We have

$$\mu_A(x) = f(x, \sigma, c) = e^{\frac{-(x-c)^2}{2\sigma^2}} . \tag{8}$$

For each of the three inputs R_e, R_i and C, we attribute the Gaussian membership functions associated with the intervals' ranges extracted from the database. It is necessary to use different zones for each input due to the performance of the learning. We attribute five membership degrees to the inputs Re and C and three to the input R_i. For each input the corresponding number of MF that leads to a high accuracy and that avoids the overlapping for the hole database represents the type of the meat.

For the fuzzy logic of freshness classification, the major difference with the fuzzy logic of muscle classification is that we do the learning for each muscle separately in order to determine the aging that corresponds to its freshness. The three model parameters R_e, R_i and C are defined as inputs and affected the trapezoidal MF after several trials.

The fuzzy logic classification results show an accuracy of 100 %. The result of each muscle is in the close interval of the corresponding value; for example, for the veal SM, the accepted result that corresponds to this muscle should be in the range from 2.5 to 3.5.

The fuzzy logic classification for noisy inputs reaches the same results as the used inputs. In the diagnosis process, the choice of confidence areas is decisive in order to support the learning process. We select and define the period of time such as:
- fresh, between day 2 and day 3;
- edible, between day 4 and day 6;
- critical, between day 7 and day 10;
- dangerous, more than day 11.

According to the defined periods, the accuracy is 84.62 % for the beef LD, 92.31 % for the beef RA, 100 % for the veal SM and 61.54 % for the veal LD.

4 Conclusion

Feature extraction using model parameters leads to a high accuracy. Suitable data for meat diagnosis are able to correctly classify muscle types and states. Fuzzy logic allows for an easy method of classification. It correctly classifies the inputs to the corresponding classes for the first step of meat type with a recognition index equal to 100 %.

For the freshness classification, fuzzy logic reaches a recognition index equal to 84.62 % for the beef LD, 92.31 % for the beef RA, 100 % for the veal SM and 61.54 % for the veal LD.

Fuzzy logic has many advantages which can be beneficial in the monitoring process of meat quality. These advantages are especially the consideration of noisy inputs, the tolerance that leads to a compact program according to the conception procedure and finally the simplicity.

Bibliography

[1] L. Guerrero, I. Gobantes, M.A Oliver, J. Arnau, M. D. Guardia, J. Elvira, P. Riu, N. Grebol, and J. M. Monfort, “Green hams electrical impedance spectroscopy (EIS) measures and pastiness prediction of dry cured hams”, Meat Science, vol. 66, no. 2, pp. 289–294, 2004.

[2] L. H. Negri, P. Bertemes-Filho, and A. S. Paterno, “Computational intelligence algorithms for bio-impedance-based classification of biological material”, In: 5th European Conference of the IFMBE, IFMBE Proceedings, vol. 37, pp. 1229–1232, 2011.

[3] P. Bertemes-Filho, L. H. Negri, and A. S. Paterno, "Detection of bovine milk adulterants using bioimpedance measurements and artificial neural network", In: 5th European Conference of the IFMBE, IFMBE Proceedings, vol. 37, pp. 1275–1278, 2011.
[4] M. Guermazi, O. Kanoun, and N. Derbel, "Reduction of anisotropy influence and contacting effects in in-vitro bioimpedance measurements", Journal of Physics Conference Series, vol. 434, pp. 1–4, 2013.
[5] M. Guermazi, O. Kanoun, and N. Derbel, "Investigation of long time beef and veal meat behaviour by bio-impedance spectroscopy for meat monitoring", IEEE Sensors Journal, vol. 14, no. 10, pp. 3624–3630, 2014.
[6] V. Yu. Musatov, V. V. Sysoev, M. Sommer, and I. Kiselev, "Assessment of meat freshness with metal oxide sensor microarray electronic nose: A practical approach", Sensors and Actuators B, Chemical, vol. 144, no. 1, pp. 99–103, 2010.
[7] T. Eklöv, G. Johansson, F. Winquist, and I. Lundström, "Monitoring sausage fermentation using an electronic nose", Journal of the Science of Food and Agriculture, vol. 76, no. 4, pp. 525–532, 1998.
[8] J. A. Balejko, N. Zbigniew, and B. Edyta, "Artificial neural network as the tool in prediction rheological features of raw minced meat", ACTA Scientiarum Polonorum. Technologia Alimentaria, vol. 11, no. 3, pp. 273–281, 2012.
[9] I. T. Jolliffe, "Principal Component Analysis", 2nd edition, Springers Series in Statistics, Springer, 2002.
[10] V. Prasath, N. Lakshmi, M. Nathiya, N. Bharathan, and N. P. Neetha, "A Survey on the applications of fuzzy logic in medical diagnosis", International Journal of Scientific & Engineering Research, vol. 4, no. 4, pp. 1199–1203, 2013.
[11] A. A. E. Saleh, S. E. Barakat, and A. A. E. Awad, "A fuzzy decision support system for management of breast cancer", International Journal of Advanced Computer Science and Applications, vol. 2, no. 3, pp. 34–40, 2011.
[12] K. C. Latha, B. Madhu, S. Ayesha, R. Ramya, R. Sridhar, and S. Balasubramanian, "Visualization of risk in breast cancer using fuzzy logic in Matlab environment", International Journal of Computational Intelligence Techniques, vol. 4, no. 1, pp. 114–117, 2013.
[13] H. J. Zimmermann, "Fuzzy Set Theory and Its Applications", ISBN 978-94-015-7951-3, London, 1991.
[14] A. M. Aziz, "Effects of fuzzy membership function shapes on clustering performance in multisensor-multitarget data fusion systems", In: IEEE International Conference on Fuzzy Systems, pp. 1839–1844, Korea, 2009.
[15] E. H. Mamdani and S. Assilian, "An experiment in linguistic synthesis with a fuzzy logic controller". International Journal of Man-Machine Studies, vol. 7, no. 1, pp. 1–13, 1975.
[16] E. H. Mamdani, "Advances in the linguistic synthesis of fuzzy controllers", International Journal of Man-Machine Studies, vol. 8, no. 6, pp. 669–678, 1976.
[17] T. Takagi and M. Sugeno, "Fuzzy identification of systems and its applications to modeling and control", IEEE Transactions on Systems, Man and Cybernetics, vol. SMC-15, no. 1, pp. 116–132, 1985.
[18] F. Zhang, "Feature Extraction for the Meat's Freshness Characterization Using the Method of Principal Component Analysis", Master of Science, TU Chemnitz, 2014.

Bei Fragen zur Produktsicherheit wenden Sie sich bitte an:
If you have any questions regarding product safety,
please contact:

Walter de Gruyter GmbH
Genthiner Straße 13
10785 Berlin
productsafety@degruyterbrill.com